SPE Petroleum Engineering Certification and PE License Exam Reference Guide

SPE Petroleum Engineering Certification and PE License Exam Reference Guide, 6th Edition

Ali Ghalambor, PhD, PE

Society of Petroleum Engineers

Disclaimer

This book was prepared by members of the Society of Petroleum Engineers and their well-qualified colleagues from material published in the recognized technical literature and from their own individual experience and expertise. While the material presented is believed to be based on sound technical knowledge, neither the Society of Petroleum Engineers nor any of the authors or editors herein provide a warranty either expressed or implied in its application. Correspondingly, the discussion of materials, methods, or techniques that may be covered by patents implies no freedom to use such materials, methods, or techniques without permission through appropriate licensing. Nothing described within this book should be construed to lessen the need to apply sound engineering judgment nor to carefully apply accepted engineering practices in the design, implementation, or application of the techniques described herein.

ISBN 978-1-61399-977-6

6th Edition

Society of Petroleum Engineers
222 Palisades Creek Drive
Richardson, TX 75080-2040 USA

https://www.spe.org/store
books@spe.org
1.972.952.9393

Table of Contents

Foreword

The Society of Petroleum Engineers (SPE) has a vision to "enable the global oil and gas E&P industry to share technical knowledge needed to meet the world's energy needs in a safe and environmentally responsible manner." One way of achieving this vision is by sustaining the competency, professionalism, impartiality, and integrity of the personnel within the industry. SPE has responded to this challenge by establishing the SPE Professional Certification Exam (SPEC), which offers members a vehicle to develop their technical competencies and skills across the entire field of petroleum engineering. The SPEC is internationally recognized and represents a high standard of knowledge in different areas of petroleum engineering via an exam that includes engineering fundamentals and complex practical problems.

The SPEC has been offered internationally now for 13 years and is complemented by an SPE short course that gives candidates insight into the range of topics that the exam will cover and the style of questions that they will face. Initially, no specific course manual existed; in most cases, the SPE *Petroleum Engineering Handbook* series was the main source of reference for the course.

In the summer of 2008, the SPE Engineering Professionalism Committee discussed the idea of writing a book that could be used as a single reference for the SPEC and exam review course. The initial concept of the book was that of a one-stop, go-to reference for future oil and gas industry professionals, with all major concepts, equations, charts, tables, and formulas between its covers. This soon evolved into a "Quick Reference Guide for Petroleum Engineers," but because the primary intended use for the book was as a reference for the SPEC and US PE Exam, it finally evolved into the guide you are holding today.

The guide has been written for a wide range of audiences and, therefore, will have many applications and uses. It will be of value to university students, recent graduates, and young professionals within the oil and gas industry and academia. However, it is also largely intended for use by experienced professionals who are working on day-to-day projects and require access to a broader scope of petroleum engineering than what falls immediately within their specific areas of expertise.

Along with being of great value before and during examinations, the guide will also be of great use in the workplace. This guide complements the *Petroleum Engineering Handbook* series by summarizing all of the concepts in a single volume. As a result, there is no need to carry a suitcase full of books on every assignment. The guide is expected to become commonplace in every department and on every desk, platform, or rig in the oil and gas industry. Additionally, the guide is anticipated to be the first-stop reference when oil industry professionals are faced with any upstream or downstream problem.

With this in mind, the *SPE Petroleum Engineering Certification and PE License Exam Reference Guide* was written in a way that will allow oil industry professionals to apply a formula or equation (that may not be at the forefront of their minds) to their daily processes and procedures without having to cross-reference other texts. The fact that the guide is largely aimed at professionals who have been in the industry for some time allows the user to be familiar with the concepts behind the procedures, so there is no need for a real textbook-style explanation behind their derivation.

In tune with the SPE vision, daily use of the guide by working engineers will increase professional standards and knowledge sharing, thus creating an industry that "meets the world's energy needs in a safe and environmentally responsible manner."

Dr. Mohammed Razik Shaikh, SPEC

Acknowledgments

The SPE Petroleum Professional Certification Subcommittee gratefully acknowledges the contributions of the author of the ***SPE Petroleum Engineering Certification and PE License Exam Reference Guide***, Ali Ghalambor, and the SPE Engineering Professionalism Committee.

The Subcommittee recognizes past and present committee members for their encouragement in preparing this guide. The Subcommittee also recognizes the review effort by the SPE Petroleum Engineering Certification and US Engineering Registration Subcommittees. The author acknowledges the assistance of Mr. Foag Haeri in literature search and gathering of information.

SPE extends its appreciation to the SPE Engineering Registration Committee for their efforts in revising the guide in 2020. Specific thanks go to Samuel Cappo, Jarrod Sparks, Paul Lammers, Chris Chamblee, Mark Fisk, Weldon Ransbarger, Lucas Moore, Neal Howard, Eric Robertson, Farrukh Hamza, George Stutz, Jared Clark, David Gaudin, Steven Tkach, and other volunteer contributors.

Chapter 1

Reservoir Engineering

1.1 Volume Calculations

Original Oil in Place in Volumetric Undersaturated Oil Reservoirs

Volumetric Method

Above Bubblepoint Pressure

$$N = \frac{7,758 \times A \times h \times \phi \times (1 - S_{wi})}{B_{oi}}$$

7,758 Number of barrels per acre-foot, bbl/acre-ft
N Original oil in place, STB
A Area of the zone, acres
h Average net thickness of the zone, ft
ϕ Porosity, unitless
B_{oi} Oil formation volume factor at initial reservoir pressure, bbl/STB
S_{wi} Water saturation at initial reservoir conditions, unitless

Below Bubblepoint Pressure

$$N = \frac{7,758 \times A \times h \times \phi \times (1 - S_{wi} - S_g)}{B_o}$$

7,758 Number of barrels per acre-foot, bbl/acre-ft
N Original oil in place, STB
A Area of the zone, acres
h Average net thickness of the zone, ft
ϕ Porosity, unitless
B_o Oil formation volume factor, bbl/STB

S_{wi} Water saturation at initial reservoir conditions, unitless

S_g Gas saturation, unitless

Material Balance Method Without Water Influx

Above Bubblepoint Pressure

$$N = \frac{N_p B_o}{B_{oi}\left(\dfrac{c_o S_o + c_w S_{wi} + c_f}{1 - S_{wi}}\right)\Delta p}$$

N Original oil in place, STB
N_p Cumulative oil produced, STB
B_o Oil formation volume factor, bbl/STB
B_{oi} Oil formation volume factor at initial reservoir pressure, bbl/STB
Δp Change in volumetric reservoir pressure, psi
c_o Oil compressibility, psi^{-1}
c_w Water compressibility, psi^{-1}
c_f Formation compressibility, psi^{-1}
S_o Oil saturation, unitless
S_{wi} Water saturation at initial reservoir conditions, unitless

Oil Formation Volume Factor

$$B_o = B_{ob}\,\text{EXP}\left[-c_o(p - p_b)\right]$$

B_o Oil formation volume factor, bbl/STB
B_{ob} Oil formation volume factor at bubblepoint pressure, bbl/STB
c_o Coefficient of compressibility, psi^{-1}
p Pressure above bubblepoint, psi
p_b Pressure at bubblepoint, psi

Below Bubblepoint Pressure

$$N = \frac{N_p[B_t + (R_p - R_{si})B_g]}{B_t - B_{ti}}$$

N Original oil in place, STB
N_p Cumulative oil produced, STB
B_t Two-phase formation volume factor, bbl/STB $= B_o + B_g\left(R_{si} - R_{so}\right)$
B_{ti} Two-phase formation volume factor at initial reservoir pressure, bbl/STB
B_g Gas formation volume factor, bbl/scf
R_p Cumulative produced gas/oil ratio, scf/STB
R_{si} Solution gas/oil ratio at initial reservoir pressure, scf/STB
R_{so} Solution gas/oil ratio, scf/STB
B_o Oil formation volume factor, bbl/STB

Recovery Factor

$$RF = \frac{N_p}{N} = \frac{(B_t - B_{ti})}{\left[B_t + \left(R_p - R_{si}\right)B_g\right]}$$

RF Recovery factor, fraction
N_p Cumulative oil produced, STB

N Original oil in place, STB
B_t Two-phase formation volume factor, bbl/STB $= B_o + B_g \left(R_{si} - R_{so} \right)$
B_{ti} Two-phase formation volume factor at initial reservoir pressure, bbl/STB
B_g Gas formation volume factor, bbl/scf
R_p Cumulative produced gas/oil ratio, scf/STB
R_{si} Solution gas/oil ratio at initial reservoir pressure, scf/STB
R_{so} Solution gas/oil ratio, scf/STB
B_o Oil formation volume factor, bbl/STB

Original Oil in Place in Undersaturated Oil Reservoirs With Water Influx

Volumetric Method

$$N = \frac{7,758 \times A \times h \times \phi \times (1 - S_{wi} - S_{or})}{B_{oi}}$$

7,758   Number of barrels per acre-foot, bbl/acre-ft
N Original oil in place, STB
A Area of the zone, acres
h Average net thickness of the zone, ft
ϕ Porosity, unitless
B_{oi} Oil formation volume factor at initial reservoir pressure, bbl/STB
S_{wi} Water saturation at initial reservoir conditions, unitless
S_{or} Residual oil saturation, unitless

Material Balance Method

$$N = \frac{N_p \left[B_t + \left(R_p - R_{si} \right) B_g \right] - W_e + B_w W_p}{B_t - B_{ti} + B_{ti} \left(\dfrac{c_w S_{wi} + c_f}{1 - S_{wi}} \right) \Delta \overline{p}}$$

N Initial oil in place, STB
N_p Cumulative oil produced, STB
B_t Two-phase formation volume factor, bbl/STB $= B_o + B_g \left(R_{si} - R_{so} \right)$
B_{ti} Two-phase formation volume factor at initial reservoir pressure, bbl/STB
B_w Water formation volume factor, bbl/STB
B_g Gas formation volume factor, bbl/scf
$\Delta \overline{p}$ Change in reservoir pressure, psi
c_w Water compressibility, psi^{-1}
c_f Formation compressibility, psi^{-1}
W_p Cumulative water produced, STB
W_e Water influx, bbl
R_p Cumulative produced gas/oil ratio, scf/STB
R_{si} Solution gas/oil ratio at initial reservoir pressure, scf/STB
S_{wi} Water saturation at initial reservoir conditions, unitless
B_o Oil formation volume factor, bbl/STB
R_{so} Solution gas/oil ratio, scf/STB

Oil Unit Recovery Factor (RF)
(Water drive with no appreciable decline in reservoir pressure)

$$RF = \frac{N_p}{N} = \frac{7,758 \times \phi \times (1 - S_{wi} - S_{or})}{B_{oi}} \,[=]\, STB / acre\text{-}ft$$

Oil Recovery Efficiency (RE)

(Water drive with no appreciable decline in reservoir pressure)

$$RE = 100 \frac{(1 - S_{wi} - S_{or})}{1 - S_{wi}} \,[=]\, \%$$

N	Original oil in place, STB
ϕ	Porosity, unitless
B_{oi}	Oil formation volume factor at initial reservoir pressure, bbl/STB
S_{wi}	Water saturation at initial reservoir conditions, unitless
S_{or}	Residual oil saturation, unitless
RF	Oil unit recovery factor, STB/acre-ft
RE	Oil recovery efficiency, %

Original Oil in Place in Saturated Oil Reservoirs

Volumetric Method

$$N = \frac{7,758 \times A_{oz} \times h_{oz} \times \phi_{oz} \times (1 - S_{wioz})}{B_{oi}}$$

7,758	Number of barrels per acre-foot, bbl/acre-ft
N	Original oil in place, STB
A_{oz}	Area of the oil zone, acres
h_{oz}	Average net thickness of the oil zone, ft
ϕ_{oz}	Average porosity in the oil zone, unitless
B_{oi}	Oil formation volume factor at initial reservoir pressure, bbl/STB
S_{wioz}	Initial average connate water saturation in the oil zone, unitless

Material Balance Method (with Water Influx)

$$N = \frac{N_p \left[B_t + (R_p - R_{si}) B_g \right] - W_e + B_w W_p}{B_t - B_{ti} + \dfrac{mB_{ti}}{B_{gi}} (B_g - B_{gi})}$$

$$m = \frac{GB_{gi}}{NB_{oi}}$$

N	Initial oil in place, STB
N_p	Cumulative oil produced, STB
B_o	Oil formation volume factor, bbl/STB
B_{oi}	Oil formation volume factor at initial reservoir pressure, bbl/STB
B_t	Two-phase formation volume factor, bbl/STB $= B_o + B_g (R_{si} - R_{so})$
B_{ti}	Two-phase formation volume factor at initial reservoir pressure, bbl/STB
B_w	Water formation volume factor, bbl/STB
B_g	Gas formation volume factor, bbl/scf
B_{gi}	Gas formation volume factor at initial reservoir pressure, bbl/scf
W_p	Cumulative water produced, STB
W_e	Water influx, bbl
R_p	Cumulative produced gas/oil ratio, scf/STB
R_{si}	Solution gas/oil ratio at initial reservoir pressure, scf/STB
R_{so}	Solution gas/oil ratio, scf/STB
m	Ratio of initial reservoir free gas volume to initial reservoir oil volume, unitless
G	Original gas in place, scf

Original Gas in Place in Volumetric Dry Gas, Wet Gas, and Retrograde Gas Condensate Reservoirs

Volumetric Method

$$G = \frac{7,758 \times A \times h \times \phi \times (1 - S_{wi})}{B_{gi}}$$

Alternatively, if B_{gi} is defined as cf/scf where cf is ft³,

$$G = \frac{43,560 \times A \times h \times \phi \times (1 - S_{wi})}{B_{gi}}$$

7,758	Number of barrels per acre-foot, bbl/acre-ft
43,560	Number of cubic feet per acre-foot, cf/acre-ft
G	Original gas in place, Mscf (or scf if alternative equation with B_{gi} as cf/scf)
A	Area of the zone, acres
h	Average net thickness of the zone, ft
ϕ	Porosity, unitless
B_{gi}	Gas formation volume factor at initial reservoir pressure, bbl/Mscf (or cf/scf for alternative equation)
S_{wi}	Water saturation at initial reservoir conditions, unitless

Applying Bulk Reservoir Volume (V_b)

$$G = \frac{43,560 \times V_b \times \phi \times (1 - S_{wi})}{B_{gi}}$$

Applying Reservoir Pore Volume (V_p)

$$G = \frac{43,560 \times V_p \times (1 - S_{wi})}{B_{gi}}$$

G	Original gas in place, scf
V_b	Bulk reservoir volume, acre-ft $= A \times h$
V_p	Reservoir pore volume, acre-ft $= A \times h \times \phi = V_b \times \phi$
ϕ	Porosity, unitless
S_{wi}	Water saturation at initial reservoir conditions, unitless
B_{gi}	Gas formation volume factor at initial reservoir pressure, cf/scf

Material Balance Method

$$\frac{p}{z} = \frac{p_i}{z_i}\left(1 - \frac{G_p}{G}\right)$$

Recovery Factor Using "B_g" Terms

$$RF = \frac{G_p}{G} = \frac{(B_g - B_{gi})}{B_g}$$

Recovery Factor Using "p/z" Terms

$$RF = \frac{G_p}{G} = \frac{\left(\dfrac{p_i}{z_i} - \dfrac{p}{z}\right)}{\dfrac{p_i}{z_i}}$$

RF	Recovery factor, fraction
G	Original gas in place, Mscf
G_p	Cumulative gas produced, Mscf
p	Reservoir pressure (current or abandonment conditions), psia
p_i	Initial reservoir pressure, psia
z	Gas compressibility factor (current or abandonment conditions), unitless
z_i	Gas compressibility factor at initial reservoir pressure, unitless
B_{gi}	Gas formation volume factor at initial reservoir pressure, bbl/Mscf
B_g	Gas formation volume factor (current or abandonment conditions), bbl/Mscf

Gas Formation Volume Factor

$$B_g = \frac{p_{sc} z T}{T_{sc} p}$$

$$B_g = \frac{0.00504 z T}{p} \; [=] \; \text{bbl/scf}$$

Alternatively, if B_g is defined as cf/scf where cf is ft^3,

$$B_g = \frac{0.02829 z T}{p} \; [=] \; \text{cf/scf}$$

B_g	Gas formation volume factor, bbl/scf (or cf/scf for alternative equation)
p_{sc}	Pressure at standard conditions, or pressure base, psia = 14.7 psia
T_{sc}	Temperature at standard conditions, °R (°R = °F + 460) = 60°F = 520°R
z	Gas compressibility factor or gas deviation factor, unitless
T	Reservoir temperature, °R
p	Reservoir pressure, psia

Pseudoreduced Pressure and Temperature

$$p_{pr} = \frac{p}{p_{pc}}$$

$$T_{pr} = \frac{T}{T_{pc}}$$

p_{pr}	Pseudoreduced pressure, unitless
p	Pressure, psia
p_{pc}	Pseudocritical pressure, psia
T_{pr}	Pseudoreduced temperature, unitless
T	Temperature, °R (°R = °F + 460)
T_{pc}	Pseudocritical temperature, °R

Specific Gravity of a Gas

$$\gamma_g = \frac{\rho_g}{\rho_{air}} = \frac{M_g}{M_{air}} = \frac{M_g}{28.97} \quad \text{(assumes gas and air obey the ideal-gas law)}$$

also $\gamma_g = \dfrac{M_a}{28.97}$

$$M_a = \sum_j y_j M_j$$

γ_g Specific gravity of a gas, unitless

ρ_g Density of a gas, lbm/ft^3

ρ_{air} Density of air, lbm/ft^3

M_g Molecular weight of a gas, lbm/lbm-mol

M_{air} Molecular weight of air (= 28.97 lbm/lbm-mol)

M_a Apparent molecular weight of a gas mixture, lbm/lbm-mol

γ_j Mole fraction of gas component j in a gas mixture, fraction

M_j Molecular weight of gas component j in a gas mixture, lbm/lbm-mol

Specific Gravity of a Reservoir Gas for a One-Stage Separation System

$$\gamma_g = \frac{R_1 \gamma_1 + 4{,}602 \gamma_o}{R_1 + 133{,}316 \gamma_o / M_o}$$

$$M_o = \frac{5{,}954}{\gamma_{API} - 8.811}$$

$$M_o = \frac{42.43 \gamma_o}{1.008 - \gamma_o}$$

$$\gamma_o = \frac{141.5}{\gamma_{API} + 131.5}$$

$$\rho_o = \frac{(62.4 \gamma_o + 0.0136 \gamma_g \times R_s)}{B_o}$$

γ_g Specific gravity of reservoir gas, unitless

R_1 Primary (high-pressure) separator gas to stock-tank liquid ratio, scf/STB

R_s Separator gas to oil ratio, scf/STB

γ_1 Specific gravity of primary separator gas, unitless (air = 1.0)

γ_{API} Specific gravity of stock-tank hydrocarbon liquid in °API

γ_o Specific gravity of the liquid hydrocarbons, unitless (water = 1.0)

M_o Molecular weight of stock-tank liquid (condensate), lbm/lbm-mol

Specific Gravity of a Reservoir Gas for a Three-Stage Separation System

$$\gamma_g = \frac{R_1 \gamma_1 + 4{,}602 \gamma_o + R_2 \gamma_2 + R_3 \gamma_3}{R_1 + (133{,}316 \gamma_o / M_o) + R_2 + R_3}$$

γ_2 Specific gravity of secondary separator gas, unitless

γ_3 Specific gravity of stock-tank gas, unitless

R_2 Secondary (low-pressure) separator gas to stock-tank liquid ratio, scf/STB

R_3 Stock-tank gas to stock-tank liquid ratio, scf/STB

Original Gas in Place in Gas Reservoirs With Water Influx

Material Balance Method

$$G = \frac{G_p B_g - W_e + B_w W_p}{B_g - B_{gi}}$$

G Initial gas in place, Mscf

G_p Cumulative gas produced, Mscf

B_w Water formation volume factor, bbl/STB

B_g Gas formation volume factor, bbl/Mscf

B_{gi} Gas formation volume factor at initial reservoir pressure, bbl/Mscf

W_p Cumulative water produced, STB

W_e Water influx, bbl

Gas Unit Recovery Factor (RF)

$$RF = 43,560 \times \phi \times \left(\frac{1-S_{wi}}{B_{gi}} - \frac{S_{gr}}{B_{ga}} \right) [=] \text{scf / acre-ft}$$

Gas Recovery Efficiency (RE)

$$RE = \frac{100 \left(\frac{1-S_{wi}}{B_{gi}} - \frac{S_{gr}}{B_{ga}} \right)}{\left(\frac{1-S_{wi}}{B_{gi}} \right)} [=] \%$$

RF Gas unit recovery factor, scf/acre-ft
RE Gas recovery efficiency, %
ϕ Porosity, unitless
S_{wi} Water saturation at initial reservoir conditions, unitless
S_{gr} Residual gas saturation, unitless
B_{gi} Gas formation volume factor at initial reservoir pressure, bbl/scf or cf/scf
B_{ga} Gas formation volume factor at abandonment reservoir pressure, bbl/scf or cf/scf

Material Balance Expressed as a Linear Equation (Havlena and Odeh 1963)

$$F = NE_t + W_e B_w$$

$$F = N \left(E_o + mE_g + E_{f,w} \right) + W_e B_w$$

$$F = N_p \left[B_o + \left(R_p - R_{so} \right) B_g \right] + W_p B_w$$

$$E_o = \left(B_o - B_{oi} \right) + \left(R_{soi} - R_{so} \right) B_g$$

$$E_g = B_{oi} \left(\frac{B_g}{B_{gi}} - 1 \right)$$

$$E_{f,w} = (1+m) B_{oi} \left(\frac{c_w S_{wi} + c_f}{1 - S_{wi}} \right) \Delta p$$

$$E_t = E_o + mE_g + E_{f,w}$$

F Underground withdrawal, bbl
E_o Oil and solution gas expansion, bbl/STB
E_g Gas cap expansion, bbl/STB
$E_{f,w}$ Hydrocarbon space reduction, bbl/STB
E_t Total expansion, bbl/STB
N Initial oil in place, STB
N_p Cumulative oil produced, STB
W_e Cumulative water influx from the aquifer into the reservoir, STB
W_p Cumulative water produced, STB
B_o Oil formation volume factor, bbl/STB
B_g Gas formation volume factor, bbl/scf
B_w Water formation volume factor, bbl/STB
B_{oi} Oil formation volume factor at initial reservoir pressure, bbl/STB
B_{gi} Gas formation volume factor at initial reservoir pressure, bbl/scf
Δp Change in reservoir pressure, psi
c_w Water compressibility, psi^{-1}
c_f Formation compressibility, psi^{-1}
S_{wi} Water saturation at initial reservoir conditions, unitless

R_p Cumulative produced gas/oil ratio, scf/STB

R_{soi} Initial solution gas/oil ratio, scf/STB

R_{so} Solution gas/oil ratio, scf/STB

m Ratio of initial reservoir free gas volume to initial reservoir oil volume, unitless

Formation Compressibility (Newman Correlations)

Isothermal Compressibility

$$c = -\frac{1}{V_\phi}\left(\frac{dV_\phi}{dp}\right)$$

Consolidated Sandstones Under Hydrostatic Pressure

$$c_f = \frac{97.3200(10)^{-6}}{(1+55.8721\phi)^{1.42859}} \quad \text{(where } 0.02 < \phi < 0.23, \; c_f \pm 2.6\%)$$

Limestone Formations Under Hydrostatic Pressure

$$c_f = \frac{0.853531}{\left[1+2.47664(10)^6\phi\right]^{0.92990}} \quad \text{(where } 0.02 < \phi < 0.33, \; c_f \pm 11.6\%)$$

c Isothermal compressibility, psi^{-1}

V_ϕ Rock pore space volume (units for volume cancel out in equation)

p Pressure, psi

c_f Formation compressibility of porous rock, psi^{-1}

ϕ Porosity, unitless

Reservoir Bulk Volume

Simpson's Rule (valid for odd-numbered layers only)

n = odd only (most accurate method)

$$V_b = \frac{h}{3}(A_1 + 4A_2 + 2A_3 + 4A_4 + 2A_5 + \ldots + 4A_{n-1} + A_n) + \frac{h_n}{3}A_n$$

Trapezoidal Rule (valid for all numbered layers)

n = even or odd (upper area at least 1/2 of lower)

$$V_b = \frac{h}{2}(A_1 + 2A_2 + 2A_3 + 2A_4 + \ldots + 2A_{n-1} + A_n) + \frac{h_n}{2}A_n$$

Pyramid Rule (valid for each layer only)

Note: For the Pyramid Rule, A_n is the area enclosed by the lower isopach line and A_{n+1} is the area enclosed by the upper isopach line.

$$\Delta V_b = \frac{h}{3}\left(A_n + A_{n+1} + \sqrt{A_n A_{n+1}}\right)$$

V_b Reservoir bulk volume, acre-ft

ΔV_b Bulk volume for single or individual layer, acre-ft

h Interval between isopach lines or common height of layers excluding top layer, ft (also called contour interval)

h_n Height of top layer, ft

n Number of layers, unitless

A_1 Area of bottom-most interval, acres

A_2 Area of second-from-bottom interval, acres

$A_{3,4,5}$ Area of designated interval counting from bottom, acres

A_{n-1} Area of second-to-last or second-from-top interval, acres

A_n Area of top-most interval, acres

1.2 Drive Mechanisms

Depletion Drive Index (DDI), or Solution Gas Drive Index (SGDI)

$$DDI = \frac{N\left(B_t - B_{ti}\right)}{N_p\left[B_t + \left(R_p - R_{soi}\right)B_g\right]}$$

Segregation (Gas Cap) Drive Index (SDI), or Gas Cap Drive Index (GCDI)

$$SDI = \frac{G\left(B_g - B_{gi}\right)}{N_p\left[B_t + \left(R_p - R_{soi}\right)B_g\right]}$$

Water Drive Index (WDI)

$$WDI = \frac{W_e - W_p B_w}{N_p\left[B_t + \left(R_p - R_{soi}\right)B_g\right]}$$

Expansion Drive Index (EDI), also referred to as Formation and Connate Water Compressibility Index (CDI) or Pore Volume Contraction Index (PVCI)

$$EDI = \frac{N B_{oi}\left(1+m\right)\left(\dfrac{c_w S_{wi} + c_f}{1 - S_{wi}}\right)\Delta p}{N_p\left[B_t + \left(R_p - R_{soi}\right)B_g\right]}$$

$$DDI + SDI + WDI + EDI = 1$$

N Initial oil in place, STB

N_p Cumulative oil produced, STB

G Initial gas in place, scf

W_e Water influx into reservoir, bbl

W_p Cumulative water produced, STB

B_t Two-phase formation volume factor, bbl/STB, $= B_o + B_g\left(R_{soi} - R_{so}\right)$

B_o Oil formation volume factor, bbl/STB

R_p Cumulative produced gas/oil ratio, scf/STB

R_{soi} Initial solution gas/oil ratio, scf/STB

R_{so} Solution gas/oil ratio, scf/STB

B_{ti} Initial two-phase formation volume factor, bbl/STB, $= B_{oi}$

B_{oi} Initial oil formation volume factor, bbl/STB

B_g Gas formation volume factor, bbl/scf

B_{gi} Initial gas formation volume factor, bbl/scf

B_w Water formation volume factor, bbl/STB

c_w Water compressibility, psi^{-1}

c_f Formation compressibility, psi^{-1}

S_{wi} Water saturation at initial reservoir conditions, unitless

Δp Change in reservoir pressure, psi

m Ratio of initial reservoir free gas volume to initial reservoir oil volume, unitless

Solution Gas Drive Mechanism

Above Bubblepoint Pressure

$$N_p B_o = N B_{oi} \left[c_o + \frac{\left(c_w S_{wi} + c_f \right)}{1 - S_{wi}} \right] \Delta p$$

$$N = \frac{N_p B_o}{B_{oi} \left[\dfrac{\left(B_o - B_{oi} \right)}{B_{oi}} + \dfrac{\left(c_w S_{wi} + c_f \right)}{1 - S_{wi}} \Delta p \right]}$$

Below Bubblepoint Pressure

$$N_p \left[B_o + \left(R_p - R_{so} \right) B_g \right] = N \left[\left(B_o - B_{oi} \right) + \left(R_{soi} - R_{so} \right) B_g \right]$$

$$N = \frac{N_p \left[B_o + \left(R_p - R_{so} \right) B_g \right]}{\left[\left(B_o - B_{oi} \right) + \left(R_{soi} - R_{so} \right) B_g \right]}$$

Gas Cap Drive Mechanism

$$N_p \left[B_o + \left(R_p - R_{so} \right) B_g \right] = N B_{oi} \left[\frac{\left(B_o - B_{oi} \right) + \left(R_{soi} - R_{so} \right) B_g}{B_{oi}} + m \left(\frac{B_g}{B_{gi}} - 1 \right) \right]$$

$$N = \frac{N_p \left[B_o + \left(R_{soi} - R_{so} \right) B_g \right]}{B_{oi} \left[\dfrac{\left(B_o - B_{oi} \right) + \left(R_{soi} - R_{so} \right) B_g}{B_{oi}} + m \left(\dfrac{B_g}{B_{gi}} - 1 \right) \right]}$$

Water Drive Mechanism

$$W_e = \left(c_w + c_f \right) W_i \Delta p$$

N	Initial oil in place, STB
N_p	Cumulative oil produced, STB
B_o	Oil formation volume factor, bbl/STB
B_g	Gas formation volume factor, bbl/scf
B_{oi}	Initial oil formation volume factor, bbl/STB
B_{gi}	Initial gas formation volume factor, bbl/scf
Δp	Change in reservoir pressure, psi
c_o	Oil compressibility, psi^{-1}
c_w	Water compressibility, psi^{-1}
c_f	Formation compressibility, psi^{-1}
S_{wi}	Water saturation at initial reservoir conditions, unitless
R_p	Cumulative produced gas/oil ratio, scf/STB
R_{soi}	Initial solution gas/oil ratio, scf/STB
R_{so}	Solution gas/oil ratio, scf/STB
W_e	Water influx into reservoir, bbl
W_i	Initial volume of water, bbl
m	Ratio of initial reservoir free gas volume to initial reservoir oil volume, unitless

1.3 Stages of Production

Darcy's Law

$$u = -\frac{k}{\mu}\frac{dp}{dL}$$

u Velocity, ft/sec
k Permeability, md
μ Viscosity, cp
p Pressure, psia
L Length, ft

Steady-State Linear Flow of Incompressible Fluids

$$q = \frac{0.001127\ k\ A\ (p_1 - p_2)}{\mu\ L}$$

q Flow rate, B/D
k Absolute permeability, md
μ Viscosity, cp
L Length, ft
A Cross-sectional area of flow, ft²
p_1 Upstream pressure, psia
p_2 Downstream pressure, psia

Steady-State Linear Flow of Slightly Compressible Fluids

$$q_{ref} = \left(\frac{0.001127\ k\ A}{\mu\ c\ L}\right) \ln\left[\frac{1 + c\left(p_{ref} - p_2\right)}{1 + c\left(p_{ref} - p_1\right)}\right]$$

q_{ref} Flow rate at a reference pressure p_{ref}, B/D
p_1 Upstream pressure, psia
p_2 Downstream pressure, psia
k Absolute permeability, md
μ Viscosity, cp
c Average liquid compressibility, psi^{-1}
L Length, ft
A Cross-sectional area of flow, ft²
p_{ref} Reference pressure, psia = 14.7 psia at standard conditions

Steady-State Linear Flow of Compressible Fluids

$$Q_{sc} = \frac{0.003164\ T_{sc}\ A\ k\ \left(p_1^2 - p_2^2\right)}{p_{sc}\ T\ L\ z\ \mu_g}$$

Q_{sc} Gas flow rate at standard conditions, scf/D
p_1 Upstream pressure, psia
p_2 Downstream pressure, psia
k Permeability, md
T Reservoir temperature, °R
T_{sc} Temperature at standard conditions, °R (°R = °F + 460), 60°F = 520°R
p_{sc} Pressure at standard conditions, or pressure base, psia = 14.7 psia
μ_g Gas viscosity, cp
A Cross-sectional area of flow, ft²
L Total length of the linear system, ft
z Gas compressibility factor, unitless

Steady-State Radial Flow of Incompressible Fluids

$$Q_o = \frac{0.00708\ k\ h\left(p_e - p_{wf}\right)}{\mu_o\ B_o\left[\ln\left(r_e/r_w\right)+s\right]}$$

Q_o Oil flow rate, STB/D
p_e External pressure, psia
p_{wf} Bottomhole flowing pressure, psia
k Permeability, md
μ_o Oil viscosity, cp
B_o Oil formation volume factor, bbl/STB
h Average net thickness of the zone, ft
r_e External or drainage radius, ft
r_w Effective wellbore radius, ft
s Skin factor, unitless

Steady-State Radial Flow of Slightly Compressible Fluids

$$q_{ref} = \frac{0.00708kh}{\mu c\ \ln\left(r_e/r_w\right)}\ln\left[\frac{1+c\left(p_{ref}-p_e\right)}{1+c\left(p_{ref}-p_{wf}\right)}\right]$$

or

$$Q_o = \frac{0.00708kh}{\mu_o B_o c_o\left[\ln\left(r_e/r_w\right)+s\right]}\ln\left[\frac{1+c_o\left(14.7-p_e\right)}{1+c_o\left(14.7-p_{wf}\right)}\right]$$

q_{ref} Flow rate at a reference pressure p_{ref}, B/D
Q_o Oil flow rate, STB/D
p_{ref} Reference pressure, psia = 14.7 psia at standard conditions
14.7   Pressure at standard conditions, psia, or pressure base = 14.7 psia
p_e External pressure, psia
p_{wf} Bottomhole flowing pressure, psia
k Permeability, md
μ Viscosity, cp
μ_o Oil viscosity, cp
B_o Oil formation volume factor, bbl/STB
h Average net thickness of the zone, ft
r_e External or drainage radius, ft
r_w Effective wellbore radius, ft
c Average liquid compressibility, psi^{-1}
c_o Oil compressibility coefficient, psi^{-1}
s Skin factor, unitless

Steady-State Radial Flow of Compressible Fluids

$$Q_g = \frac{k\ h\left(\psi_e - \psi_w\right)}{1,422\ T\ln\left(r_e/r_w\right)}$$

$$Q_{g(P<2000)} = \frac{k\ h\left(p_e^2 - p_{wf}^2\right)}{1,422\ T(\mu_g z)_{avg}\ln\left(r_e/r_w\right)}$$

Q_g Gas flow rate, Mscf/D
$Q_{g(P<2000)}$ Gas flow rate calculated at pressures (P) less than 2,000 psia, Mscf/D
k Permeability, md
h Average net thickness of the zone, ft
T Reservoir temperature, °R

r_e	External or drainage radius, ft
r_w	Effective wellbore radius, ft
μ_g	Gas viscosity, cp
z	Gas compressibility factor, unitless
$(\mu_g z)_{avg}$	Product of gas-viscosity and gas-compressibility factor calculated at the average pressure between p_e and p_{wf}, cp
ψ_e	Real-gas pseudopressure as evaluated from 0 to p_e, psi²/cp
ψ_w	Real-gas pseudopressure as evaluated from 0 to p_{wf}, psi²/cp
p_e	External pressure, psia
p_{wf}	Bottomhole flowing pressure, psia
P	Pressure, psia

Unsteady-State (Transient) Radial Flow of Slightly Compressible Fluids (Diffusivity Equation)

$$\frac{\partial^2 p}{\partial r^2} + \frac{1}{r}\frac{\partial p}{\partial r} = \frac{\phi\,\mu\,c_t}{0.006328\,k}\frac{\partial p}{\partial t}$$

k	Permeability, md
r	Radial position, ft
p	Pressure, psia
c_t	Total compressibility, psi⁻¹ $\quad c_t = c_o S_o + c_w S_w + c_g S_g + c_f$
t	Time, days
ϕ	Porosity, unitless
μ	Viscosity, cp
c_o	Oil compressibility, psi⁻¹
c_w	Water compressibility, psi⁻¹
c_g	Gas compressibility, psi⁻¹
c_f	Formation compressibility, psi⁻¹
S_o	Oil saturation, unitless
S_w	Water saturation, unitless
S_g	Gas saturation, unitless

Unsteady-State (Transient) Radial Flow of Compressible Fluids (Diffusivity Equation)

$$\frac{\partial^2 m(p)}{\partial r^2} + \frac{1}{r}\frac{\partial m(p)}{\partial r} = \frac{\phi\,\mu\,c_t}{0.000264\,k}\frac{\partial m(p)}{\partial t}$$

$$m\left(p_{wf}\right) = m\left(p_i\right) - 57{,}895.3\left(\frac{p_{sc}}{T_{sc}}\right)\left(\frac{Q_g T}{kh}\right)\left[\log\left(\frac{kt}{\phi\mu_i c_{ti} r_w^2}\right) - 3.23\right]$$

$m(p)$	Real-gas psuedopressure, psia²/cp
p_i	Initial reservoir pressure, psia
p_{wf}	Bottomhole flowing pressure, psia
Q_g	Gas flow rate, Mscf/D
t	Time, hr
k	Permeability, md
p_{sc}	Pressure at standard conditions, or pressure base, psia = 14.7 psia
T_{sc}	Temperature at standard conditions, °R (°R = °F + 460), 60°F = 520°R
T	Reservoir temperature, °R
r_w	Effective wellbore radius, ft
h	Average net thickness of the zone, ft
μ_i	Gas viscosity at the initial pressure, cp
c_{ti}	Total compressibility coefficient at p_i, psi⁻¹
ϕ	Porosity, unitless

Pseudosteady-State Radial Flow of Slightly Compressible Fluids

$$Q = \frac{0.00708\ k\ h\ (\bar{p}_r - p_{wf})}{\mu B \left[\ln(r_e / r_w) - 0.75 + s \right]}$$

$$J = \frac{Q}{\bar{p}_r - p_{wf}} = \frac{kh}{141.2 B \mu \left[\ln(r_e / r_w) - 0.75 + s \right]}$$

Q	Flow rate, STB/D
$\bar{p}_r$	Average reservoir pressure, psia, $= \sqrt{(p_r^2 + p_{wf}^2)/2}$ for $p < 2{,}000$
p_{wf}	Bottomhole flowing pressure, psia
k	Permeability, md
h	Average net thickness of the zone, ft
μ	Oil viscosity, cp
B	Formation volume factor, bbl/STB
r_e	External or drainage radius, ft
r_w	Effective wellbore radius, ft
s	Skin factor, unitless
J	Productivity index, STB/D/psi
p_r	External or reservoir pressure, psia

Pseudosteady-State Radial Flow of Compressible Fluids

$$Q_g = \frac{k\ h\left(\bar{p}_r^{\,2} - p_{wf}^{\,2}\right)}{1{,}422\ T\ \bar{\mu}\ \bar{z}\left[\ln(r_e / r_w) - 0.75 + s \right]}$$

Q_g	Gas flow rate, Mscf/D
$\bar{p}_r$	Average reservoir pressure, psia
p_{wf}	Bottomhole flowing pressure, psia
k	Permeability, md
h	Average net thickness of the zone, ft
T	Reservoir temperature, °R
$\bar{\mu}$	Average gas viscosity, cp
$\bar{z}$	Average gas compressibility factor, unitless
r_e	External or drainage radius, ft
r_w	Effective wellbore radius, ft
s	Skin factor, unitless

The E_i-Function Solution to the Diffusivity Equation (Constant Rate)

$$p_{wf} = p_i - \frac{162.6\ Q_o\ B_o\ \mu_o}{k\ h}\left[\log\left(\frac{kt}{\phi \mu_o c_t r_w^2} \right) - 3.23 + 0.87s \right]$$

p_{wf}	Bottomhole flowing pressure, psia
p_i	Initial reservoir pressure, psia
k	Permeability, md
t	Time, hr $\ t > 9.48 \times 10^4 \left(\phi \mu_o c_t r_w^2 / k \right)$
h	Average net thickness of the zone, ft
μ_o	Oil viscosity, cp
B_o	Oil formation volume factor, bbl/STB
Q_o	Oil flow rate, STB/D
r_w	Effective wellbore radius, ft
c_t	Total compressibility, psi^{-1} $\ c_t = c_o S_o + c_w S_w + c_g S_g + c_f$
ϕ	Porosity, unitless
s	Skin factor, unitless
c_o	Oil compressibility, psi^{-1}

c_w	Water compressibility, psi^{-1}
c_g	Gas compressibility, psi^{-1}
c_f	Formation compressibility, psi^{-1}
S_o	Oil saturation, unitless
S_w	Water saturation, unitless
S_g	Gas saturation, unitless

1.4 Well Performance

Pressure Drawdown Analysis (or Constant Terminal-Rate Solutions)

$$k = \frac{162.6 q \mu B}{mh}$$

$$s = 1.151 \left(\frac{p_i - p_{1hr}}{m} - \log \frac{k}{\phi \mu c_t r_w^2} + 3.23 \right)$$

$$t_{wbs} \approx \frac{(200,000 + 12,000 s) C_s}{(kh / \mu)}$$

$$C_s = \frac{qB}{24} \frac{\Delta t}{\Delta p}$$

Δt and Δp are values read from a point on the unit-slope line on the log-log plot. A less-acceptable alternative is to use the actual mechanical properties of the well:

For a Well With a Rising Liquid/Gas Interface in the Wellbore

$$C_s = 25.65 \frac{A_{wb}}{\rho}$$

For a Wellbore Containing Only Single-Phase Fluid (Liquid or Gas)

$$C_s = c_{wb} V_{wb}$$

Reservoir Pore Volume

$$V_p = \frac{-0.234 qB}{c_t \left(\frac{\partial p_{wf}}{\partial t} \right)}, \text{ft}^3$$

$\left(\dfrac{\partial p_{wf}}{\partial t} \right)$ is the slope of the straight line p_{wf} vs. t plot on Cartesian graph paper.

Transient Period

E_i form (at any r):

$$p(r,t) = p_i + \frac{70.6q\ B\ \mu}{k\ h} E_i \left(-\frac{948 \phi \mu c_t r^2}{kt} \right)$$

The mathematical function, E_i, is known as the exponential integral and is defined as

$$E_i(-x) = -\int_x^{\infty} \frac{e^{-u} du}{u} = \left[\ln x - \frac{x}{1!} + \frac{x^2}{2(2!)} - \frac{x^3}{3(3!)} + \text{etc.} \right]$$

where, in the case $x = \dfrac{948 \phi \mu c_t r^2}{kt}$,

the E_i-function solution is an accurate approximation of more-exact solutions to the diffusivity equation for time $3.79 \times 10^5 \phi \mu c_t r_w^2 / k < t < 948 \phi \mu c_t r_e^2 / k$.

For $x < 0.01$, the E_i function can be approximated by

$E_i(-x) = \ln(1.781x)$.

Log form of $p(r,t)$ equation is valid for $x < 0.01$ $\left(\text{or } \dfrac{948\phi\mu c_t r^2}{kt} < 0.01 \right)$:

$$p(r,t) = p_i - m\left[\log\left(\frac{kt}{\phi\mu c_t r^2}\right) - 3.23\right]$$

For $x > 10$, the E_i function can be considered zero for practical applications in flow through porous media:

$E_i(-x) = 0$.

For $0.01 < x < 10$, the E_i functions can be determined from tables (i.e., the $E_i(-x)$ integral is calculated as a function of x over this range, and the resulting values are presented in a table).

Horner's Approximation

$$t_p = 24 \times \frac{N_p}{q_{last}}$$

$$p_i - p = -\frac{70.6\mu q_{last} B}{kh} E_i\left(-\frac{948\phi\mu c_t r^2}{kt_p}\right)$$

Pseudosteady State

$$p_{wf}(t) = p_i - m\left[\log\left(\frac{4A}{\gamma C_A r_w^2}\right) + 0.87s\right] - \frac{0.2339qBt}{c_t Ah\phi}$$

Time (hours) When Pseudosteady State Begins:

$$t = \frac{\phi\mu c_t A t_{DA}}{0.000264k}$$

t_{DA} is dependent on reservoir shape factor C_A, and can be read from the "Stabilized Conditions for $\dfrac{kt}{\phi\mu cA} >$" column in **Fig. 1.7.**

Time to Pseudoradial Flow For Fractured Wells or Wells Producing From a Linear Fracture (Lee and Wattenbarger 1996)

$$t_{prf} - \text{hrs} = \frac{11,400\phi\mu_g c_t L_f^2}{k_g}$$

Pressure Buildup Analysis

Horner Equation

$$p_{ws}(\Delta t) = p_i - m\log\left(\frac{t_p + \Delta t}{\Delta t}\right)$$

$$k = \frac{162.6q\mu B}{mh}$$

$$r_i = \left(\frac{kt}{948\phi\mu c_t}\right)^{1/2}$$

$$s = 1.151\left[\frac{(p_{1hr} - p_{w\Delta t=0})}{m} - \log\left(\frac{k}{\mu c_t \phi r_w^2}\right) + 3.23\right]$$

$$t_{wbs} \approx \frac{170,000 C_s e^{0.14s}}{(kh/\mu)}$$

$$C_s = \frac{qB}{24} \frac{\Delta t}{\Delta p}$$

Δt and Δp are values read from a point on the unit-slope line on the log-log plot.

A less-acceptable alternative is to use the actual mechanical properties of the well:

For a Well With a Rising Liquid/Gas Interface in the Wellbore

$$C_s = 25.65 \frac{A_{wb}}{\rho}$$

For a Wellbore Containing Only Single-Phase Fluid (Liquid or Gas)

$$C_s = c_{wb} V_{wb}$$

Gray's Equation for Distance From Well to Fracture

$$L = \sqrt{\frac{0.000148 k \Delta t_x}{\phi \mu c_t}}$$

Interference Testing

$$r_D = \frac{r}{r_w}$$

$$t_D = \frac{0.000264 kt}{\phi \mu c_t r_w^2}$$

$$p_D \approx \frac{1}{2}\left[\ln\left(t_D / r_D^2\right) + 0.80907\right]$$

$$\Delta p = p_i - p_r = \left(\frac{141.2 qB\mu}{kh}\right) p_D$$

Earlougher's Radius of Investigation (Earlougher 1977)

$$r_i = 0.029 \sqrt{\frac{kt}{\phi \mu c_t}}$$

p_i	Initial reservoir pressure, psia
V_p	Reservoir pore volume, ft³
p_{wf}	Wellbore flowing pressure, psia
p_{ws}	Wellbore pressure during buildup or shut-in, psia
$p_{w\Delta t = 0}$	Wellbore pressure at instant of shut-in, psia
p_{1hr}	Horner shut-in pressure, psia, at $\Delta t = 1$ hr
t	Elapsed time, hr
t_{DA}	Dimensionless time based on drainage area A, unitless
t_p	Effective producing time or pseudoproducing time, hr
ϕ	Porosity, unitless
h	Average net thickness of the zone, ft
μ	Viscosity, cp
B	Formation volume factor, RB/STB
k	Permeability, md
c_t	Total compressibility, psi⁻¹, $c_t = c_o S_o + c_w S_w + c_g S_g + c_f$
c_o	Oil compressibility, psi⁻¹
c_w	Water compressibility, psi⁻¹
c_g	Gas compressibility, psi⁻¹
c_f	Formation compressibility, psi⁻¹
S_o	Oil saturation, unitless

S_w	Water saturation, unitless
S_g	Gas saturation, unitless
r	Radius or distance from center of active well to center of observation well, ft
r_D	Dimensionless radius, unitless
r_i	Radius of investigation, ft
r_w	Effective wellbore radius, ft
r_e	External or drainage radius, ft
Δp	Pressure drawdown at the observation well, psi, $= p_i - p_r$
Δt	Shut-in time, hr
q	Flow rate, STB/D
γ	Euler's constant = 1.781
A	Drainage area, ft^2
C_A	Dietz shape factor, unitless = 31.6 for a circular reservoir
m	Absolute value of slope of middle-time line, psi/log cycle $= \dfrac{162.6qB\mu}{kh}$
s	Skin factor, unitless
C_s	Wellbore storage constant, bbl/psi
A_{wb}	Wellbore area, ft^2
ρ	Density of liquid in wellbore, lbm/ft^3
c_{wb}	Compressibility of liquid in wellbore, psi^{-1}
V_{wb}	Wellbore volume, bbl
t_{wbs}	Wellbore storage duration, hr
t_D	Dimensionless time, unitless
p_D	Dimensionless pressure, unitless
p_r	Pressure at radius r, or at observation well, psia
N_p	Cumulative oil produced from well, STB
q_{last}	Most recent oil flow rate, STB/D
p	Pressure or current reservoir pressure, psia

1.5 Secondary Recovery Processes

Interfacial tension (IFT)

$$\sigma = \frac{rh(\rho_w - \rho_a)g}{2\cos\theta}$$

σ	Interfacial tension, dynes/cm
r	Capillary tube radius, cm
h	Height of water rise in the capillary, cm
ρ_w	Water density, g/cm^3
ρ_a	Air density, g/cm^3
g	Gravity acceleration constant, 980 cm/s^2
θ	Contact angle between water and capillary tube, degrees

Capillary Pressure in a Tube

$$p_c = \frac{2\sigma\cos\theta}{r}$$

p_c	Capillary pressure, dynes/cm^2
σ	Interfacial tension between two immiscible phases, dynes/cm
θ	Contact angle or wetting angle, degrees
r	Radius of the capillary tube, cm

Effective and Relative Permeabilities

$$k_w = \frac{q_w B_w \mu_w L}{0.001127\, A_c \Delta p}$$

$$k_{rw} = \frac{k_w}{k}$$

$$k_o = \frac{q_o B_o \mu_o L}{0.001127 \, A_c \Delta p}$$

$$k_{ro} = \frac{k_o}{k}$$

k_w	Effective permeability to water, md
k_o	Effective permeability to oil, md
k_{rw}	Relative permeability to water, fraction
k_{ro}	Relative permeability to oil, fraction
k	Absolute permeability, md
q_w	Water flow rate, STB/D
q_o	Oil flow rate, STB/D
B_w	Water formation volume factor, bbl/STB
B_o	Oil formation volume factor, bbl/STB
μ_w	Water viscosity, cp
μ_o	Oil viscosity, cp
L	Length of flow path, ft
A_c	Cross-sectional area of flow, ft^2
Δp	Pressure differential, psi

Empirical Relationship Between Relative Permeabilities and Water Saturation

$$\frac{k_{ro}}{k_{rw}} = a e^{bS_w}$$

k_{ro}	Relative permeability to oil, fraction
k_{rw}	Relative permeability to water, fraction
a, b	Empirically derived contants
S_w	Water saturation, unitless

Mobility Ratio of Water to Oil

$$M_{w,o} = \frac{(\lambda_w)_{S_{or}}}{(\lambda_o)_{S_{wc}}} = \left(\frac{k_{rw}}{\mu_w}\right)_{S_{or}} \left(\frac{\mu_o}{k_{ro}}\right)_{S_{wc}}$$

$M_{w,o}$	Mobility ratio of water to oil, unitless
λ_w	Mobility of displacing fluid or water behind the front, fraction
λ_o	Mobility of displaced fluid or oil ahead of the front, fraction
S_{or}	Residual oil saturation, unitless
S_{wc}	Connate water saturation, unitless
k_{rw}	Relative permeability to water evaluated at residual oil saturation, S_{or}, fraction
k_{ro}	Relative permeability to oil evaluated at connate water saturation, S_{wc}, fraction
μ_o	Oil viscosity, cp
μ_w	Water viscosity, cp

Fractional Flow Equation for Water Displacing Oil in a Linear Horizontal System

$$f_w = \frac{1}{1 + \left(\frac{k_o}{k_w}\right)\left(\frac{\mu_w}{\mu_o}\right)} = \frac{1}{1 + \left(\frac{k_{ro}}{k_{rw}}\right)\left(\frac{\mu_w}{\mu_o}\right)}$$

f_w	Fractional flow of water (water cut), bbl/bbl, fraction
k_w	Effective permeability to water, md
k_o	Effective permeability to oil, md
k_{ro}	Relative permeability to oil, fraction

k_{rw} Relative permeability to water, fraction

μ_w Water viscosity, cp

μ_o Oil viscosity, cp

General Form of Fractional Flow Equation

$$f_w = \frac{1+\left(\dfrac{0.001127k_o A}{\mu_o q_t}\right)\left(\dfrac{\partial p_c}{\partial x}-0.433\Delta\rho\sin\alpha\right)}{1+\left(\dfrac{k_o}{k_w}\right)\left(\dfrac{\mu_w}{\mu_o}\right)}$$

f_w Fractional flow of water (water cut), bbl/bbl, fraction

k_w Effective permeability to water, md

k_o Effective permeability to oil, md

μ_w Water viscosity, cp

μ_o Oil viscosity, cp

A Cross-sectional area of flow, ft^2

$\Delta\rho$ Water-oil density differences, g/cm^3, $(\rho_w - \rho_o)$

ρ_w Water density, g/cm^3

ρ_o Oil density, g/cm^3

q_t Total flow rate, B/D, $(q_w + q_o)$

α Reservoir dip angle, degrees, positive updip (or angle measured from the horizontal to the line indicating the direction of flow)

x Distance, ft

p_c Capillary pressure, psi, $(p_o - p_w)$

p_o Oil-phase pressure, psi

p_w Water-phase pressure, psi

Frontal-Advance Equation (Buckley-Leverett)

$$x = \frac{5.615 q_t t}{\phi A}\left(\frac{df_w}{dS_w}\right)_{S_w}$$

5.615   Number of cubic feet per barrel, ft^3/bbl

x Distance traveled by a fixed saturation in time, t, ft

q_t Total flow rate, B/D, $(q_w + q_o)$

q_w Water flow rate, B/D

q_o Oil flow rate, B/D

t Time interval, days

ϕ Porosity, unitless

A Cross-sectional area of flow, ft^2

S_w Water saturation, unitless

$\left(\dfrac{df_w}{dS_w}\right)_{S_w}$ Slope of fractional flow curve at S_w

Breakthrough Time

$$t_{bt} = \frac{LA\phi}{5.615 q_i\left(\dfrac{df_w}{dS_w}\right)_{S_{wf}}}$$

5.615   Number of cubic feet per barrel, ft^3/bbl

t_{bt} Time to water breakthrough at the producer, days

q_i Injection rate, B/D

L Linear distance from injection well to production well, ft

A Cross-sectional area of flow, ft²

ϕ Porosity, unitless

S_{wf} Water saturation at the producing well or flood front, unitless

$\left(\dfrac{df_w}{dS_w}\right)_{S_{wf}}$ Slope of the main tangent line at S_{wf}

Recoverable Oil at Breakthrough Time

$$N_{P_{bt}} = \frac{LA\phi\left(\overline{S_{wf}} - S_{iw}\right)}{5.615}$$

5.615 Number of cubic feet per barrel, ft³/bbl

$N_{P_{bt}}$ Recoverable oil at breakthrough time, STB

L Linear distance from injection well to production well, ft

A Cross-sectional area of flow, ft²

ϕ Porosity, unitless

$\overline{S_{wf}}$ Average water saturation at breakthrough, unitless

S_{iw} Initial water saturation prior to flood, unitless

Modified Hall-Plot Equation for Monitoring the Performance of the Injector

$$\sum\left[\left(p_{inj} - \overline{p}\right) \times \Delta t\right] = m_H W_i$$

Δt Injection time, days

p_{inj} Bottomhole injection pressure, psia

$\overline{p}$ Average reservoir pressure, psia

W_i Cumulative water injection volume, bbl

m_H Slope of Hall plot, $= \dfrac{141.2\,\mu_w\,B_w\left[\ln\left(r_e/r_w\right) + s\right]}{k_w h}$

k_w Effective permeability to water, md

r_e External or drainage radius, ft

r_w Effective wellbore radius, ft

μ_w Water viscosity, cp

h Average net thickness of the zone, ft

B_w Water formation volume factor, bbl/STB

s Skin factor, unitless

Critical Rate for Stable Displacement

To predict the critical velocity required to propagate a stable interface through a linear system in which gravity forces dominate with piston-like displacement and neglected capillary pressure (p_c) effects:

$$q_c = \frac{\left(4.9 \times 10^{-4}\right) k k_{rw} A(\rho_w - \rho_o)\sin\alpha}{\mu_w (M - 1)}$$

q_c Critical injection-rate threshold for stable displacement, RB/D

ρ_o Oil density, g/cm³

ρ_w Water density, g/cm³

α Reservoir dip angle, degrees

k_{rw} Relative permeability to water, fraction

k Absolute permeability, md

A Cross-sectional area of flow, ft²

μ_w Water viscosity, cp

M Mobility ratio of water to oil, unitless

Displacement Efficiency of a Waterflood

$$E_D = 1 - \frac{S_{or}/B_{oa}}{S_{oi}/B_{oi}}$$

E_D Displacement efficiency of a waterflood, fraction
S_{or} Residual oil saturation at the end of the flood, unitless
S_{oi} Initial oil saturation at the start of the flood, unitless
B_{oa} Oil formation volume factor at the waterflood pressure, bbl/STB or RB/STB
B_{oi} Oil formation volume factor at the start of the flood, bbl/STB or RB/STB

Overall Waterflood Oil Recovery Efficiency

$$E_R = E_D E_I E_A$$

E_R Waterflood oil recovery efficiency, fraction
E_D Unit displacement efficiency, fraction
E_I Vertical displacement efficiency, fraction
E_A Areal displacement efficiency, fraction

Displacement Efficiency of a Waterflood (Assuming a Constant Oil Formation Volume Factor During the Life of the Flood)

$$E_D = \frac{\overline{S}_w - S_{wi} - S_{gi}}{1 - S_{wi} - S_{gi}}$$

E_D Displacement efficiency of a waterflood, fraction
$\overline{S}_w$ Average water saturation in the swept area, unitless
S_{wi} Initial water saturation at the start of the flood, unitless
S_{gi} Initial gas saturation at the start of the flood, unitless

Areal Sweep Efficiency

Before Breakthrough

$$E_A = \frac{W_i}{(V_p)(\overline{S_{wf}} - S_{wi})}$$

E_A Fraction of the area that has been swept to an average water saturation of $\overline{S_{wf}}$, fraction
$\overline{S_{wf}}$ Average water saturation behind the front at breakthrough, unitless
W_i Cumulative water injected, bbl
V_p Flood pattern pore volume, bbl, $= 7,758 Ah\phi$
S_{wi} Initial water saturation at the start of the flood, unitless
A Area of the zone, acres
h Average net thickness of the zone, ft
ϕ Porosity, unitless

At Breakthrough

$$E_{A_{bt}} = 0.546 + \frac{0.0317}{M_{bt}} + \frac{0.30223}{e^{M_{bt}}} - 0.0051 M_{bt}$$

$$W_{i_{bt}} = E_{A_{bt}} V_p \left(\overline{S_{wf}} - S_{wi}\right)$$

$W_{i_{bt}}$ Cumulative water injected at breakthrough, bbl
$E_{A_{bt}}$ Areal sweep efficiency at breakthrough of the displacing fluid, fraction

M_{bt} Mobility ratio at breakthrough, unitless

V_p Flood pattern pore volume, bbl $= 7{,}758 A h \phi$

$\overline{S_{wf}}$ Average water saturation behind the front at breakthrough, unitless

S_{wi} Initial water saturation at the start of the flood, unitless

A Area of the zone, acres

h Average net thickness of the zone, ft

ϕ Porosity, unitless

After Breakthrough

$$E_A = E_{A_{bt}} + 0.274 \ln\left(\frac{W_i}{W_{ibt}}\right)$$

E_A Fraction of the area that has been swept to an average water saturation of $\overline{S_{wf}}$, fraction

$E_{A_{bt}}$ Areal sweep efficiency at breakthrough of the displacing fluid, fraction

W_i Cumulative water injected, bbl

$W_{i_{bt}}$ Cumulative water injected at breakthrough, bbl

$\overline{S_{wf}}$ Initial water saturation at the start of the flood, unitless

Water Injectivity Equation (Five-Spot Pattern Filled With Oil)

$$\left(\frac{i}{\Delta p}\right)_{\text{base}} = \frac{0.003541\, h\, k\, k_{ro}}{\mu_o\left(\ln\dfrac{d}{r_w} - 0.619\right)}$$

If including skin factors for injector (s_i) and producer (s_p):

$$\left(\frac{i}{\Delta p}\right)_{\text{base}} = \frac{0.003541\, h\, k\, k_{ro}}{\mu_o\left[\ln\dfrac{d}{r_w} + \dfrac{s_i + s_p}{2} - 0.619\right]}$$

i_{base} Base (initial) water injection rate, B/D

Δp_{base} Base (initial) pressure differential, psi, $p_{wi} - p_{wp}$

p_{wi} Injection well pressure (downhole), psia

p_{wp} Production well pressure (downhole), psia

h Average net thickness of the zone, ft

k Absolute permeability, md

k_{ro} Relative permeability to oil as evaluated at S_{wi}, fraction

S_{wi} Initial water saturation at the start of the flood, unitless

d Distance between injector and producer, ft

r_w Effective wellbore radius, ft

μ_o Oil viscosity, cp

s_i Skin factor for the injection well, unitless

s_p Skin factor for the production well, unitless

Approximate Pressure at Water-Injection Well

$$p_{wi} \approx 0.433 \rho_w D + p_{whi}$$

0.433 psi/ft = pressure gradient for fresh water

p_{wi} Injection well pressure (downhole), psi

ρ_w Water density, g/cm^3

D Well depth [in true vertical depth (TVD)], ft

p_{whi} Injection wellhead pressure, psi

Dimensionless Gravity Number

$$G = \frac{4.9 \times 10^{-4} k \ k_{rw} \ A \ (\rho_w - \rho_o)\sin\infty}{i_w \ \mu_w}$$

G Dimensionless gravity number, unitless
k Absolute permeability, md
k_{rw} Relative permeability to water as evaluated at residual oil saturation, S_{or}, fraction
A Cross-sectional area of flow, ft^2
ρ_w Water density, g/cm^3
ρ_o Oil density, g/cm^3
μ_w Water viscosity, cp
i_w Water injection rate, B/D
∞ Reservoir dip angle, degrees

Vertical Sweep Efficiency (Stiles' Method)

$$E_V = \sum_{i=1}^{j} \frac{h_i}{h_t} + \sum_{i=j+1}^{n} \frac{\frac{k_i}{k_j} h_i}{h_t}$$

E_V Vertical sweep efficiency or fraction of vertical cross section swept to S_{or}, fraction
S_{or} Residual oil saturation, unitless
i Layer number, i.e., $i = 1,2,3, \dots n$
j Last layer flooded out
n Total number of layers (noncommunicating with no crossflow between layers)
h_t Total thickness, ft
h_i Layer thickness, ft
k_i Layer permeability, md
k_j Layer permeability for the last layer flooded out, md

Fill-Up Water Volume

$$W_{if} = S_g V_p / B_w$$

W_{if} Water injection volume to fill-up, STB
S_g Gas saturation, unitless
V_p Reservoir pore volume, bbl, $= 7,758 A h \phi$
A Area of the zone, acres
h Average net thickness of the zone, ft
ϕ Porosity, unitless
B_w Water formation volume factor, bbl/STB or RB/STB

Recoverable Oil From Waterflooding

$$N_{pf} = 7,758 \phi E_t V_{sw} \left(\frac{S_{oP}}{B_{oP}} - \frac{S_{or}}{B_{or}} \right)$$

N_{pf} Recoverable oil from a waterflood, STB
ϕ Porosity, unitless
E_t Overall flood efficiency, or fraction of gross swept volume that is flushed by water during the flood, fraction (portion in flood pattern)
V_{sw} Gross swept volume of flood, acre-ft, $= Ah$
A Area of the zone, acres
h Average net thickness of the zone, ft

S_{oP}	Oil saturation at the start of the flood or after primary recovery, unitless
S_{or}	Residual oil saturation, or remaining oil saturation at abandonment in a part of the reservoir that has been flushed by water during the flood, unitless
B_{oP}	Oil formation volume factor at the start of the flood, bbl/STB or RB/STB
B_{or}	Oil formation volume factor at the waterflood pressure, bbl/STB or RB/STB

Oil Saturation at the Start of Waterflood

$$S_{oP} = \frac{(N - N_{pP}) B_{oP} (1 - S_{wc})}{N B_{oi}}$$

S_{oP}	Oil saturation at the start of the flood or after primary recovery, unitless
N	Initial oil in place, STB
N_{pP}	Cumulative oil produced at the start of the flood or after primary recovery, STB
B_{oP}	Oil formation volume factor at the start of the flood, bbl/STB or RB/STB
S_{wc}	Connate water saturation, unitless
B_{oi}	Oil formation volume factor at initial reservoir conditions, bbl/STB or RB/STB

Original Oil in Place Before Primary Production

$$N = \frac{7,758\phi (1 - S_{wc}) V_{bP}}{B_{oi}}$$

N	Initial oil in place, STB
ϕ	Porosity, unitless
S_{wc}	Connate water saturation, unitless
V_{bP}	Bulk volume, or reservoir volume drained under primary production, acre-ft; whereas V_{sw} is the reservoir volume drained under secondary recovery (e.g., waterflood) and is dependent on the waterflood pattern (i.e., $V_{sw} \leq V_{bP}$)
B_{oi}	Oil formation volume factor at initial reservoir conditions, bbl/STB or RB/STB

Flood Front Map

$$r_{ob} = \left(\frac{5.615 W_i E_{inj}}{\pi \phi h S_g} \right)^{1/2}$$

r_{ob}	Outer radius of the banked oil, ft
W_i	Cumulative water injected, bbl
S_g	Gas saturation at start of injection, unitless
E_{inj}	Layer injection efficiency (fraction of injected water volume that enters the layer where effective water-flood is taking place)
h	Average net thickness of the zone, ft
ϕ	Porosity, unitless

$$r_{wb} = r_{ob} \left(\frac{S_g}{\overline{S}_{wbt} - S_{wc}} \right)^{1/2}$$

r_{wb}	Water bank radius, ft
$\overline{S}_{wbt}$	Average water saturation behind front, unitless; for approximate calculations $\overline{S}_{wbt} \approx 1.0 - S_{orw}$
S_{wc}	Connate water saturation, unitless
S_g	Gas saturation, unitless
S_{orw}	Residual oil saturation to water injection, unitless

Waterflood Patterns

Direct Line Drive

$$q_i = \frac{0.003541\ kh\Delta p}{\mu_d \left(\ln \dfrac{a}{r_w} + 1.571 \dfrac{d}{a} - 1.838 + \dfrac{s_i + s_p}{2} \right)}, \qquad \frac{d}{a} \geq 1$$

Staggered Line Drive

$$q_i = \frac{0.003541\ kh\Delta p}{\mu_d \left(\ln \dfrac{a}{r_w} + 1.571 \dfrac{d}{a} - 1.838 + \dfrac{s_i + s_p}{2} \right)}$$

Five-Spot

$$q_i = \frac{0.003541\ kh\Delta p}{\mu_d \left(\ln \dfrac{d}{r_w} - 0.619 + \dfrac{s_i + s_p}{2} \right)}$$

$$s = \ln \left(\frac{r_s}{r_w} \right) \left(\frac{k}{k_s} - 1 \right)$$

s	Skin factor, unitless
s_i	Skin factor for the injection well, unitless
s_p	Skin factor for the production well, unitless
r_s	Skin model radius with a permeability of k_s, ft ($r_s > r_w$)
k_s	Altered permeability in immediate area of wellbore resulting from introduction of skin factor, md
q_i	Injection rate, B/D
k	Absolute permeability, md
h	Average net thickness of the zone, ft
Δp	Pressure differential, psi, $= p_{wi} - p_{wp}$
p_{wi}	Injection well pressure (downhole), psi
p_{wp}	Production well pressure (downhole), psi
μ_d	Viscosity of the displaced phase, or the fluid being displaced by the waterflood, cp
r_w	Effective wellbore radius, ft
d	Distance between a producer well and an injector well, or between a line of producers and a line of injectors, ft
a	Distance between two adjacent producer wells or two injector wells in a row, ft

Reservoir Water Cut (f_w) and Water/Oil Ratio (WOR) Relationships

Reservoir f_w – Reservoir WOR_r Relationship

$$WOR_r = \frac{f_w}{1 - f_w}$$

Reservoir f_w – Surface WOR_s Relationship

$$WOR_s = \frac{B_o f_w}{B_w (1 - f_w)}$$

Reservoir WOR_r – Surface WOR_s Relationship

$$WOR_s = WOR_r \left(\frac{B_o}{B_w} \right)$$

Surface WOR_s – Surface f_{ws} Relationship

$$f_{ws} = \frac{WOR_s}{WOR_s + 1}$$

Reservoir f_w – Surface f_{ws} Relationship

$$f_{ws} = \frac{B_o}{B_w\left(\dfrac{1}{f_w} - 1\right) + B_o}$$

WOR_s	Surface water/oil ratio, STB/STB
WOR_r	Reservoir water/oil ratio, bbl/bbl or RB/RB
f_{ws}	Surface water cut, STB/STB, fraction
f_w	Reservoir water cut, bbl/bbl or RB/RB, fraction
B_w	Water formation volume factor, bbl/STB or RB/STB
B_o	Oil formation volume factor, bbl/STB or RB/STB

Leverett j-Function

Leverett and coworkers based on the evaluation of gas/water capillary pressure data for drainage and imbibition in unconsolidated sands, proposed the following definition:

$$P_{cgw} = \sigma_{gw}\sqrt{\frac{\phi}{k}}\; j(S_w).$$

The function $j(S_w)$, defined in the equation above, is known to many as the "Leverett j-function." The j-function is obtained from experimental data by plotting $\dfrac{P_{cgw}}{\sigma_{gw}}\sqrt{\dfrac{k}{\phi}}$ against S_w.

$j(S_w)$	Leverett j-function
k	Permeability, L^2, md
P_{cgw}	Capillary pressure between gas and water phases, m/Lt^2, psi
S_w	Saturation of water
σ_{gw}	Gas/water interfacial tension, m/t^2, dyne/cm
ϕ	Porosity

1.6 Tertiary Recovery Processes

Welge Equation for the Fractional Flow of Gas

$$f_g = \frac{1 + \left(0.044 k k_{ro}\Delta\rho A\sin\alpha \Big/ q_t\mu_o\right)}{1 + 1/M} \quad (k \text{ in darcies})$$

A	Area of cross section normal to the bedding plane, ft^2
f_g	Fraction of flowing stream that is gas, fraction
k	Absolute permeability, darcies
k_{ro}	Relative permeability to oil evaluated at connate water saturation, S_{wc}, fraction
k_{rg}	Relative permeability to gas evaluated at residual oil saturation, S_{org}, fraction
M	Mobility ratio of gas to oil, $= \dfrac{k_{rg}}{k_{ro}}\dfrac{\mu_o}{\mu_g}$, unitless
q_t	Total flow rate through area A, res ft^3/D
α	Reservoir dip angle, positive downdip, degrees
$\Delta\rho$	Gas-oil density difference, lbm/ft^3, $= \rho_g - \rho_o$
ρ_g	Gas density, lbm/ft^3
ρ_o	Oil density, lbm/ft^3
μ_o	Oil viscosity, cp
μ_g	Gas viscosity, cp

Critical Rate in Displacing the Oil by Gas

$$\left(\frac{q_T}{A}\right)_{critical} = \frac{0.044 k \Delta \rho \sin \alpha}{\dfrac{\mu_o}{k_{ro}} - \dfrac{\mu_g}{k_{rg}}} \quad (k \text{ in darcies})$$

q_T	Total flow rate through area A, res ft³/D
k	Absolute permeability, darcies
k_{ro}	Relative permeability to oil, evaluated at connate water saturation, S_{wc}, fraction
k_{rg}	Relative permeability to gas evaluated at residual oil saturation, S_{org}, fraction
α	Reservoir dip angle, positive downdip, degrees
$\Delta \rho$	Gas-oil density difference, lbm/ft³, $= \rho_g - \rho_o$
ρ_g	Gas density, lbm/ft³
ρ_o	Oil density, lbm/ft³
μ_o	Oil viscosity, cp
μ_g	Gas viscosity, cp
A	Area of cross section normal to the bedding plane, ft²

Empirical Flory Equation for Polymers

$$d_p = 8 \left(M_p [\eta] \right)^{1/3}$$

d_p	Mean end-to-end distance of a polymer in solution, Angstroms (10^{-10} m)
$[\eta]$	Polymer's intrinsic viscosity, dl/g
M_p	Polymer molecular weight, g/mol

Resistance Factor of the Polymer Solution (Mobility Reduction)

$$R_f = \frac{\lambda_w}{\lambda_p}$$

R_f	Resistance factor of the polymer solution, or ratio of the mobility of a solvent to a polymer solution, unitless
λ_w	Mobility of the solvent of the polymer solution, fraction
λ_p	Mobility of the polymer solution, fraction

Residual Resistance Factor of the Polymer Solution (Permeability Reduction)

$$R_{rf} = \frac{k_b}{k_a}$$

R_{rf}	Residual resistance factor of the polymer solution, or polymer-induced permeability reduction, unitless
k_b	Brine permeability measured before polymer flooding, md
k_a	Brine permeability measured after polymer flooding, md

Polymer Retention Measurements Conversion

$$\Gamma_v = 2.7194 \, \Gamma \, (1 - \phi) \, \rho_{RG}$$

ϕ	Porosity, unitless
ρ_{RG}	Density of the reservoir rock grains (no pore space included)
Γ	Mass of polymer absorbed onto reservoir rock per unit mass of reservoir rock, µg/g
Γ_v	Mass of polymer absorbed per unit volume of reservoir rock, lbm/acre-ft

Prediction of MMP for Nitrogen or Lean Gas Injection (Firoozabadi and Aziz Correlation)

$$MMP = 9,433 - 188 \times 10^3 \left(\frac{C_{2-5}}{M_{C7+}T^{0.25}} \right) + 1,430 \times 10^3 \left(\frac{C_{2-5}}{M_{C7+}T^{0.25}} \right)^2$$

MMP Minimum miscibility pressure, psia
M_{C7+} Molecular weight of C_{7+} (Heptanes plus), lbm/lbm-mol
C_{2-5} Concentration of C_2 through C_5 including CO_2 and H_2S in the reservoir fluid, mol%
T Reservoir temperature, °F

1.7 Reserves Estimations

Exponential Decline Method for Rate

$q = q_i e^{-Dt}$ (D and t should always be on the same time basis – years, months, days, and so on.)

Exponential Decline Method for Cumulative Oil Production

$$N_p = \frac{(q_i - q)}{D}(c)$$

Exponential Decline Method for Nominal Decline Rate

$$D = -\ln(1 - D_e)$$

(D and D_e should reflect the period of time, t, between q_i and q; i.e., one year would be a D of x/yr and a D_e of y/yr)

$$D_e = \frac{q_i - q}{q_i}$$

Exponential Decline Method for Effective Decline Rate

$$D_e = 1 - e^{-D}$$

Exponential Decline Method for Life

$$t = \frac{\ln(q_i / q)}{D}$$

q Well production rate at time t, STB/D
q_i Well production rate at time 0, STB/D
D Nominal exponential decline rate, fraction/yr
D_e Effective decline rate, fraction/yr
N_p Cumulative oil production, STB
t Time, year
c Time factor, 365 D/yr (in this case), used to calculate N_p from daily production rates and annual decline rates (if used, monthly rates would be 30.4 D/m.)

Hyperbolic Decline Method for Rate

$q = q_i (1 + bD_i t)^{-\frac{1}{b}}$ (D and t should always be on the same time basis – years, months, days, for example).

Hyperbolic Decline Method for Cumulative Oil Production

$$N_p = \frac{q_i^b}{D_i(1-b)} \left(q_i^{1-b} - q^{1-b} \right)(c)$$

Hyperbolic Decline Method for Nominal Decline Rate

$$D_i = \frac{1}{b}\left[\left(1-D_{ei}\right)^{-b} - 1\right]$$

$$D_{ei} = \frac{q_i - q}{q_i}$$

$$D_t = \frac{D_i}{1 + bD_i t}$$

Hyperbolic Decline Method for Effective Decline Rate

$$D_e = 1 - e^{-D}$$

Hyperbolic Decline Method for Life

$$t = \frac{\left(q_i / q\right)^b - 1}{bD_i}$$

q Well production rate at time t, STB/D

q_i Well production rate at time 0, STB/D

D_i Initial nominal exponential decline rate ($t = 0$), fraction/yr

D_e Effective decline rate, fraction/yr

D_{ei} Initial effective decline rate, fraction/yr

D Nominal decline rate, fraction/yr

D_t Nominal decline rate at time t, fraction/yr

N_p Cumulative oil production, STB

b Hyperbolic exponent, $b > 0$, $b \neq 1$, unitless (if $b = 1$, then it is defined as Harmonic decline)

t Time, yr

c Time factor, 365 D/yr (in this case), used to calculate N_p from daily production rates and annual decline rates (if used, monthly rates would be 30.4 D/m)

Harmonic Decline Method for Rate

$$q = \frac{q_i}{1 + bD_i t} \quad \text{(D and t should always be on the same time basis – years, months, days, and so on.)}$$

Harmonic Decline Method for Cumulative Oil Production

$$N_p = \frac{q_i}{D_i} \ln \frac{q_i}{q}(c)$$

Harmonic Decline Method for Nominal Decline Rate

$$D_i = \frac{D_{ei}}{1 - D_{ei}}$$

$$D_{ei} = \frac{q_i - q}{q_i}$$

$$D_t = \frac{D_t}{1 + D_i t}$$

Harmonic Decline Method for Effective Decline Rate

$$D_e = 1 - e^{-D}$$

Harmonic Decline Method for Life

$$t = \frac{(q_i / q) - 1}{D_i}$$

q	Well production rate at time t, STB/D
q_i	Well production rate at time 0, STB/D
D_i	Initial nominal exponential decline rate ($t = 0$), fraction/yr
D_e	Effective decline rate, fraction/yr
D_{ei}	Initial effective decline rate, fraction/yr
D	Nominal decline rate, fraction/yr
D_t	Nominal decline rate at time t, fraction/yr
N_p	Cumulative oil production, STB
b	Hyperbolic exponent, $= 1$, unitless (Harmonic decline defined as $b = 1$)
t	Time, yr
c	Time factor, 365 D/yr (in this case), used to calculate N_p from daily production rates and annual decline rates (if used, monthly rates would be 30.4 D/m)

Analogy Method

$$F_{RS} = F_{RA} (\phi S_{hi})_S / (\phi S_{hi})_A$$

F_{RS}	Recovery factor, subject reservoir, STBO/acre-ft or Mscf/acre-ft
F_{RA}	Recovery factor, analogous reservoir, STBO/acre-ft or Mscf/acre-ft
ϕ	Porosity, unitless
S_{hi}	Initial hydrocarbon saturation, unitless

Volumetric Method

Initial Reserves of Oil

$$N_{Ri} = N_i E_{Ro} = \left[7{,}758 \ \phi_o (1 - S_{wo}) A_o h_{no} / B_{oi} \right] E_{Ro}$$

7,758	Number of barrels per acre-foot, bbl/acre-ft
N_{Ri}	Initial oil reserves, STB
N_i	Initial oil in place, STB
E_{Ro}	Recovery efficiency of oil, fraction
ϕ_o	Porosity in oil zone, unitless
S_{wo}	Water saturation in the oil zone, unitless
A_o	Area of the oil zone, acres
h_{no}	Average net oil pay, ft
B_{oi}	Initial oil formation volume factor, bbl/STB or RB/STB

Initial Reserves of Solution Gas (or Associated Gas)

$$G_{RSi} = G_{Si} E_{Rg} = N_i R_{si} E_{Rg}$$

G_{RSi}	Initial solution gas reserves, scf
G_{Si}	Initial solution gas in place, scf
E_{Rg}	Recovery efficiency of gas, fraction
N_i	Initial oil in place, STB
R_{si}	Initial solution gas/oil ratio, scf/STB

Initial Reserves of Gas (for Gas Caps or Nonassociated Gas)

$$G_{RFi} = G_{Fi} E_{Rg} = \left[7{,}758 \phi_g (1 - S_{wg}) A_g h_{ng} / B_{gi} \right] E_{Rg}$$

7,758	Number of barrels per acre-foot, bbl/acre-ft

7,758   Number of barrels per acre-foot, bbl/acre-ft
G_{RFi} Initial free gas reserves, Mscf
G_{Fi} Initial free gas in place, Mscf
E_{Rg} Recovery efficiency of gas, fraction
ϕ_g Porosity in gas zone, unitless
S_{wg} Water saturation in the free-gas zone, unitless
A_g Area of gas cap or gas reservoir, acres
h_{ng} Average net thickness of gas cap or gas reservoir, ft
B_{gi} Initial gas formation volume factor, bbl/Mscf or RB/Mscf

Initial Reserves of Condensate (for Gas Cap or Gas Reservoir)

$$C_{Ri} = C_i E_{Rc} = G_{Fi} R_{ci} E_{Rc}$$

C_{Ri} Initial condensate reserves, STB
C_i Initial condensate (distillate) in place, STB
E_{Rc} Recovery efficiency of condensate, fraction
G_{Fi} Initial free gas in place, MMscf (note units)
R_{ci} Initial condensate/gas ratio, STB/MMscf

1.8 Unconventional Reservoirs

Shale Gas Reserves Estimation

$$G_i = 1.359 A h_s \rho C_{gi}$$

G_i Gas in place at initial reservoir conditions, Mscf
A Area of accumulation, acres
h_s Shale thickness, ft
ρ Shale density, g/cm^3
C_{gi} Initial sorbed gas concentration or initial gas content, scf/ton

Volumetric Estimate of Gas in Place in Coalbed Reservoirs

$$G_i = Ah\left[\frac{43,560\ \phi_f\left(1-S_{wfi}\right)}{B_{gi}} + 1.359 C_{gi}\rho_c\left(1-f_a-f_m\right)\right]$$

43,560   Number of cubic feet per acre-foot, cf/acre-ft
G_i Gas in place at initial reservoir conditions, Mscf
A Area of accumulation, acres
h Net coal thickness, ft
ϕ_f Interconnected fracture (effective) porosity or cleat porosity, unitless
S_{wfi} Interconnected fracture water saturation or initial water saturation fraction in the cleats, unitless
B_{gi} Initial gas formation volume factor, ft^3/Mscf or rcf/Mscf
C_{gi} Initial sorbed gas concentration or initial gas content, scf/ton (dry, ash-free basis)
ρ_c Coal density, g/cm^3 (dry, ash-free basis)
f_a Average weight fraction of ash, fraction
f_m Average weight fraction of moisture, fraction

Langmuir Isotherm Relationship in Coalbed Reservoirs

$$G_s = G_{sL}\left[1-\left(w_a+w_{we}\right)\right]\frac{p}{p+p_L}$$

G_s Gas storage capacity, scf/ton
G_{sL} Dry, ash-free Langmuir storage capacity, scf/ton
w_a Ash content (in-situ basis), weight fraction

w_{we} Equilibrium moisture content (in-situ basis), weight fraction

p Pressure, psia

p_L Langmuir pressure, psia

Critical Desorption Pressure in Coalbed Reservoirs

$$p_c = \frac{p_L G_{ci}}{G_{sL}\left[1-\left(w_a+w_{we}\right)\right]-G_{ci}}$$

p_c Critical desorption pressure, psia

p_L Langmuir pressure, psia

G_{ci} Initial gas content (in-situ basis), scf/ton

G_{sL} Dry, ash-free Langmuir storage capacity, scf/ton

w_a Ash content (in-situ basis), weight fraction

w_{we} Equilibrium moisture content (in-situ basis), weight fraction

Gas Recovery Factor in Coalbed Reservoirs

$$f_g = 1 - \frac{G_{sL}\left[1-\left(w_a+w_{we}\right)\right]\bar{p}}{G_{ci}\left(\bar{p}+p_L\right)}$$

f_g Fractional gas recovery, fraction

G_{sL} Dry, ash-free Langmuir storage capacity, scf/ton

w_a Ash content (in-situ basis), weight fraction

w_{we} Equilibrium moisture content (in-situ basis), weight fraction

$\bar{p}$ Average reservoir pressure (at abandonment), psia

G_{ci} Initial gas content (in-situ basis), scf/ton

p_L Langmuir pressure, psia

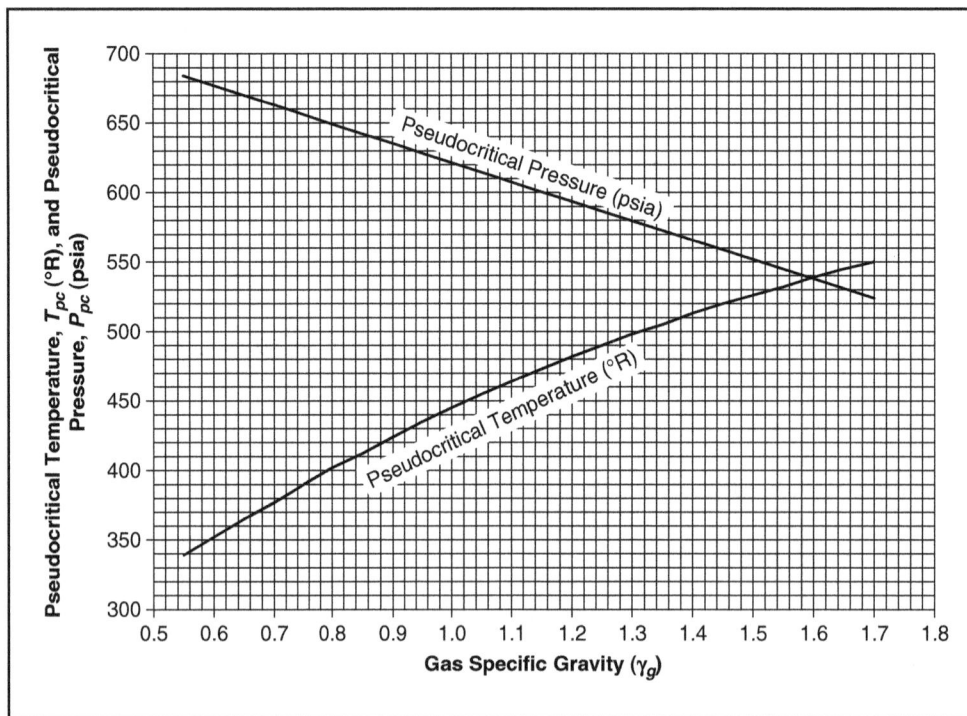

Fig. 1.1—Pseudocritical properties of natural gases (Sutton 1985).

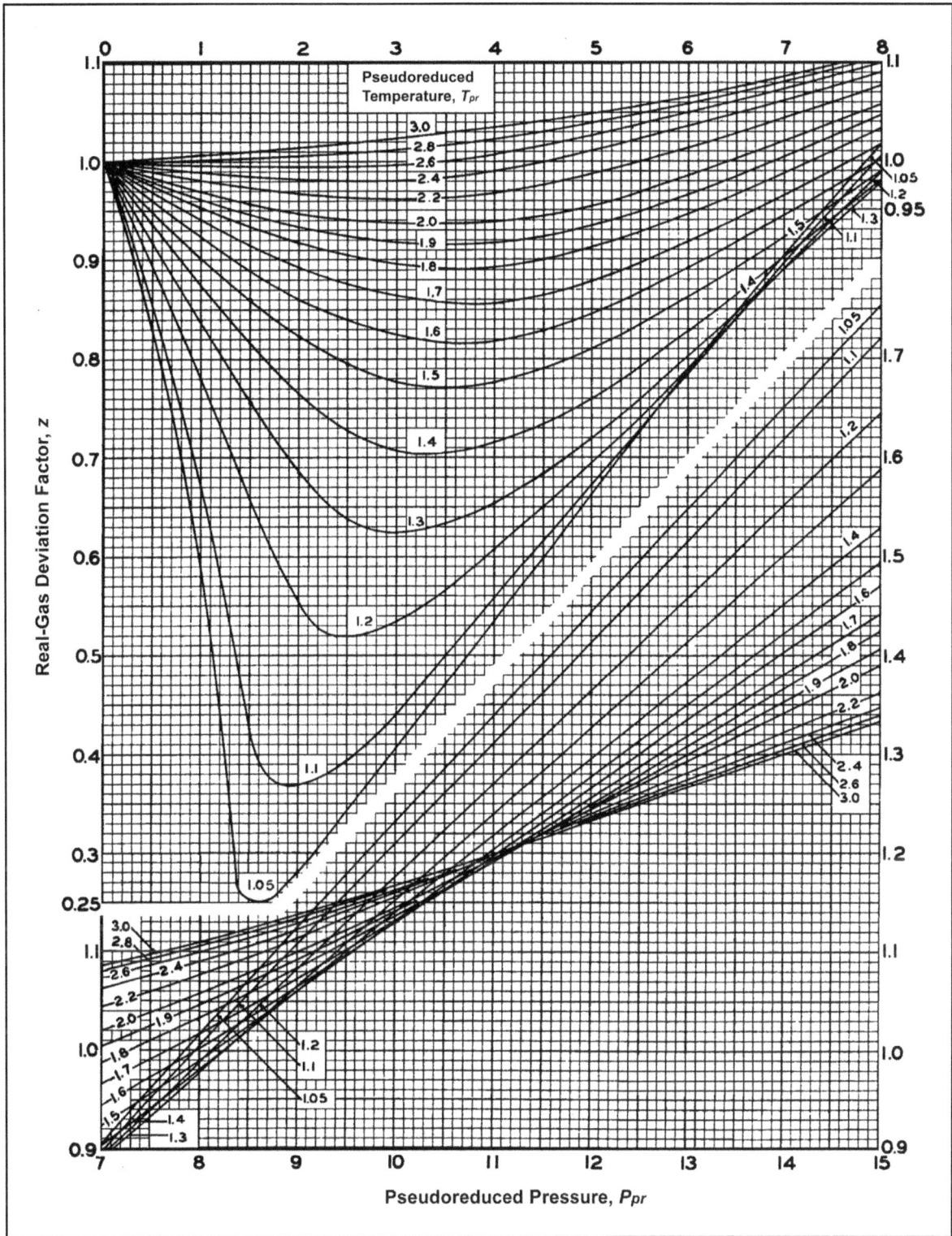

Fig. 1.2—Gas-law deviation or compressibility factors for natural gases (Standing and Katz 1942).

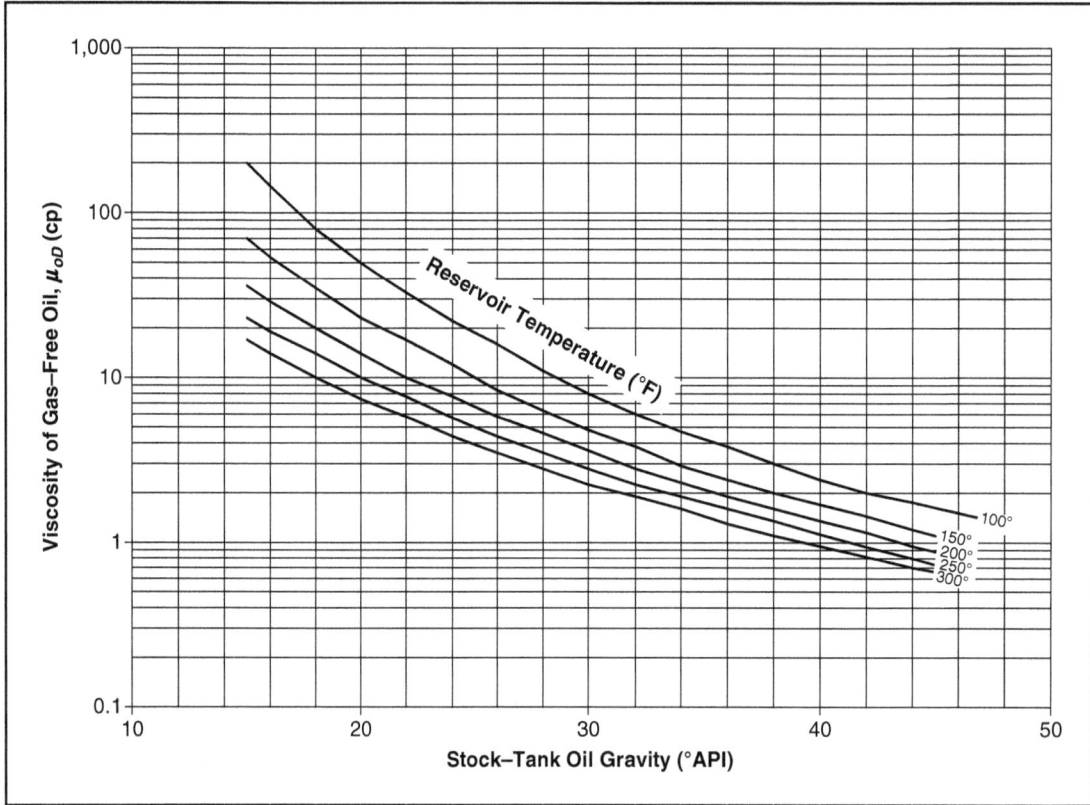

Fig. 1.3—Dead oil viscosities (McCain 1990a).

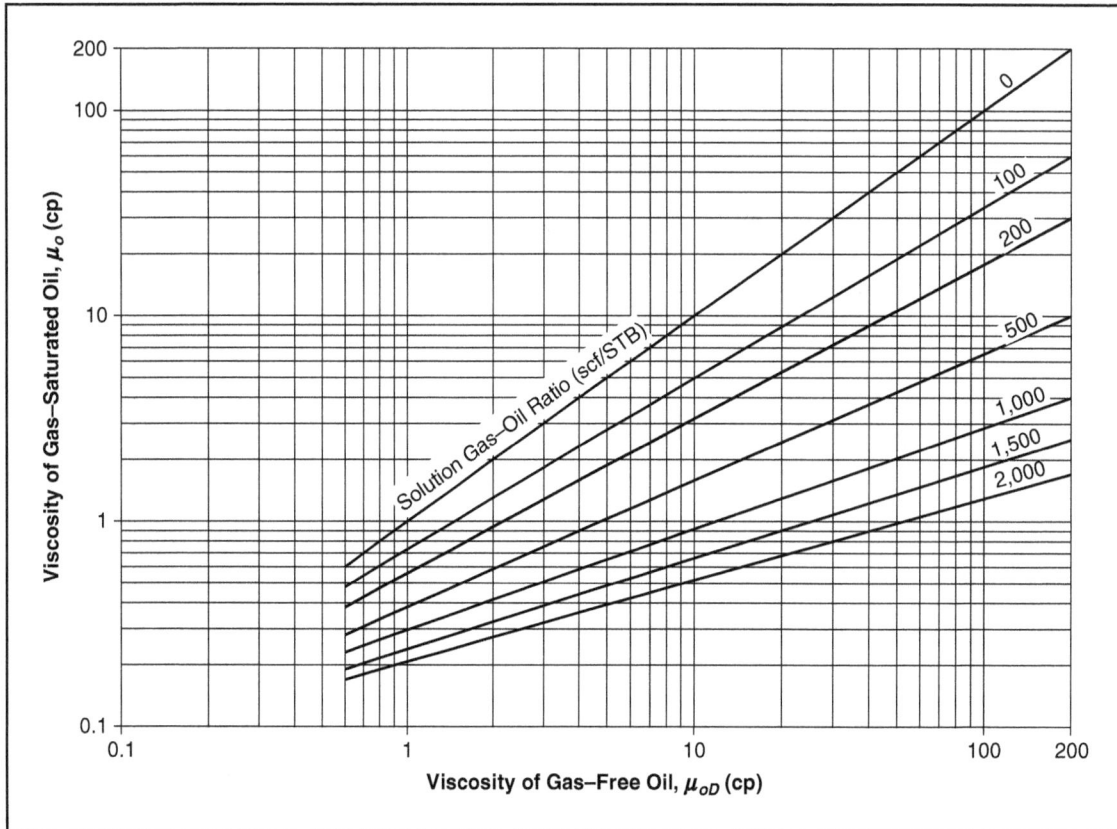

Fig. 1.4—Viscosities of saturated black oils (McCain 1990b).

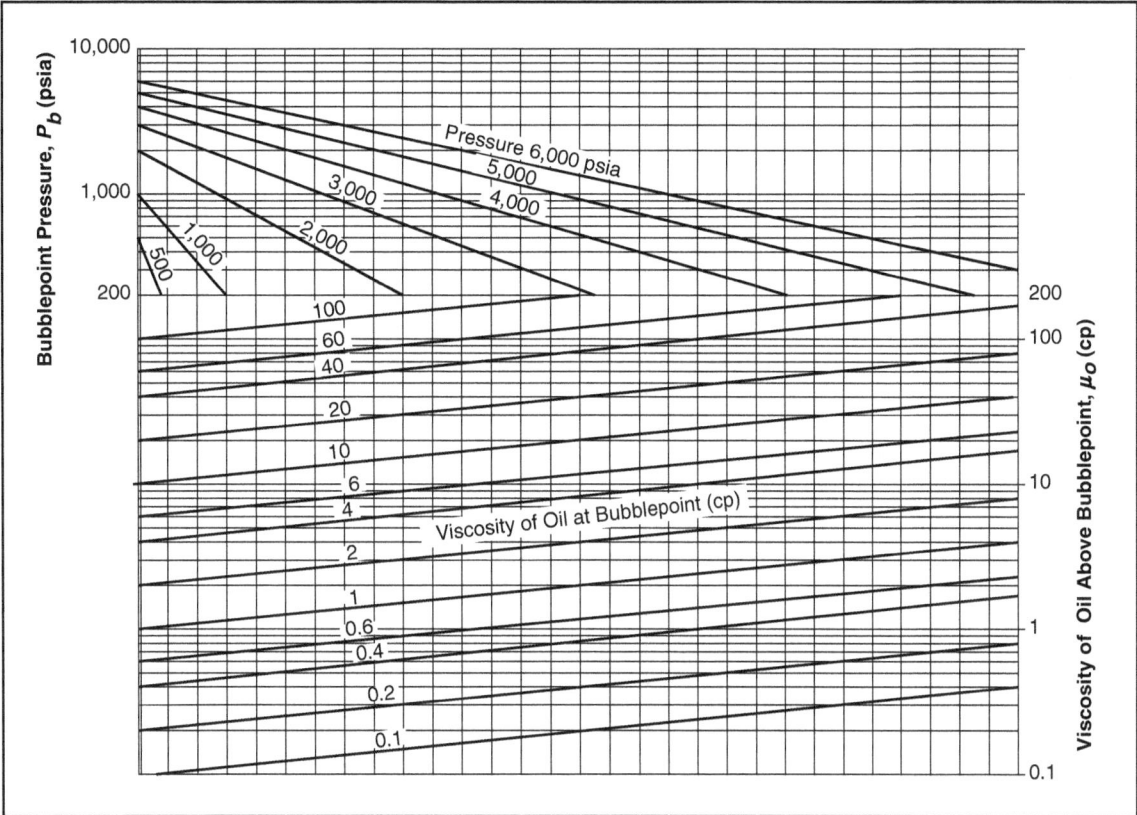

Fig. 1.5—Viscosities of undersaturated black oils (McCain 1990c).

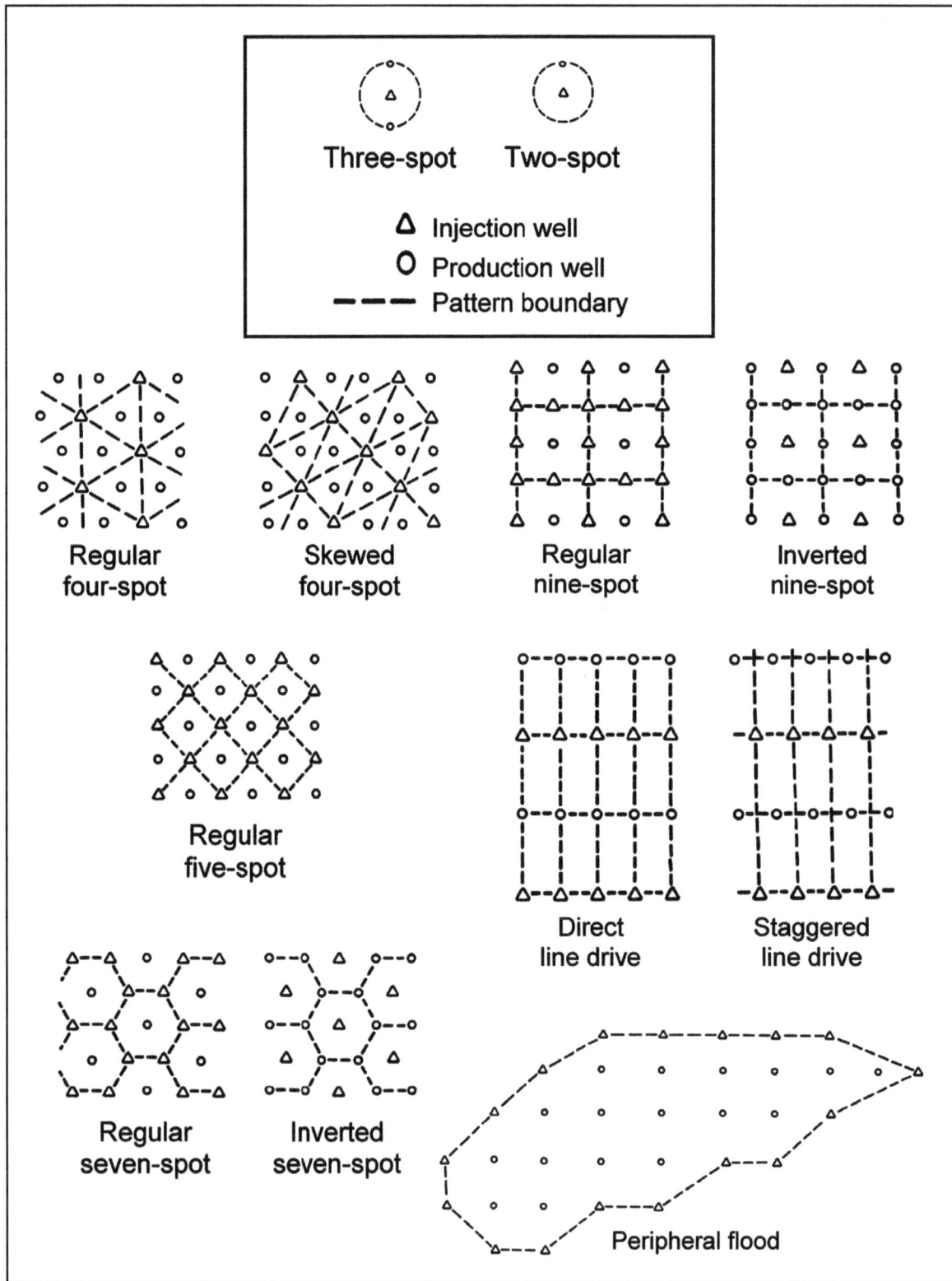

Fig. 1.6—Flooding patterns (Willhite 1986).

bounded reservoirs	$\ln C_A$	C_A	for $\frac{kt}{\phi\mu\,cA} >$		$\ln C_A$	C_A	for $\frac{kt}{\phi\mu\,cA} >$
(circle)	3.45	31.6	0.1		2.38	10.8	0.3
(square)	3.43	30.9	0.1		1.58	4.86	1.0
(hexagon)	3.45	31.6	0.1		0.73	2.07	0.8
(triangle)	3.32	27.6	0.2		1.00	2.72	0.8
(parallelogram 60°)	3.30	27.1	0.2		−1.46	0.232	2.5
(triangle 1/3)	3.12	21.9	0.4		−2.16	0.115	3.0
(rectangle 1/2)	3.12	22.6	0.2		1.22	3.39	0.6
(rectangle 1/4)	1.68	5.38	0.7		1.14	3.13	0.3
(rectangle 1/5)	0.86	2.36	0.7		−0.50	0.607	1.0
(grid)	2.56	12.9	0.6		−2.20	0.111	1.2
(grid)	1.52	4.57	0.5		−2.32	0.098	0.9
				In water-drive reservoirs	2.95	19.1	0.1
				In reservoirs of unknown production character	3.22	25	0.1

Fig. 1.7—Dietz shape factors for various geometries (Dake 1978).

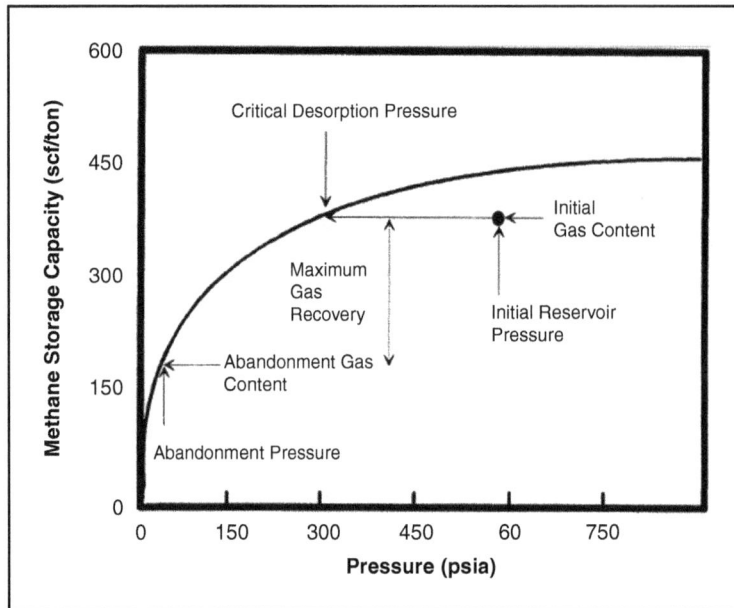

Fig. 1.8—Example sorption isotherm curve for coalbed reservoirs (Crain 2018).

Chapter 2

Drilling Engineering

2.1 General Information

Hydrostatic Pressure (HP)

$HP, psi = .052 \times$ Mud Weight, ppg $\times$ True Vertical Depth, ft

Capacity Factor (CF)

$$CF, bbl/ft = \frac{(Diameter, in.)^2}{1,029.4}$$

Open-Ended Pipe Displacement

$$Open\text{-}Ended\ Pipe\ Displacement, bbl/ft = \frac{(Pipe\ OD, in.)^2 - (Pipe\ ID, in.)^2}{1,029.4}$$

Buoyancy Factor (BF)

$$BF = \frac{\rho_{steel} - mud\ weight, ppg}{\rho_{steel}}$$

ρ_{steel} Density of steel, lbf/gal

Drillpipe Classification (remaining wall thickness)

New	1.0
Critical	.875
Premium	.8

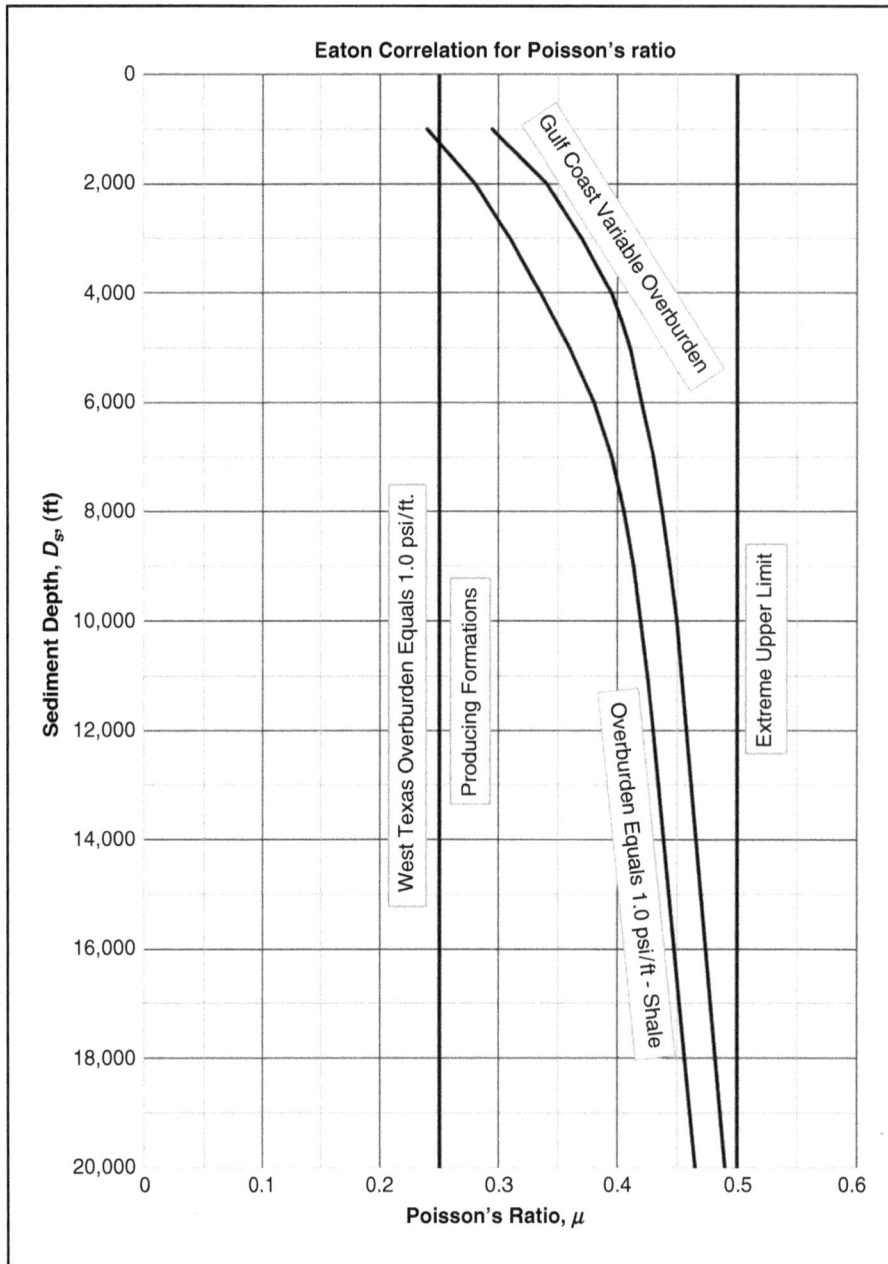

Fig. 2.1—Eaton correlation for Poisson's ratio (Bourgoyne et al. 1986a).

2.2 Geomechanics

Eaton's Method: Pore Pressure

$$P_p = S - \left(S - P_{\text{hyd}}\right)\left(R_{\text{log}} / R_{\text{n}}\right)^{1.2}$$

$$P_p = S - \left(S - P_{\text{hyd}}\right)\left(\Delta T_{\text{n}} / \Delta T_{\text{log}}\right)^{3.0}$$

P_p Pore pressure, psi

P_{hyd} Hydrostatic pressure, psi

S Total stress, psi

ΔT_{log} Measured value of sonic transit-time at a given depth, μ_{s}/ft

ΔT_{n} Normal value of sonic transit-time at a given depth, μ_{s}/ft

$R_{\log}$ Measured value of resistivity, ohm-m

R_{n} Normal value of resistivity, ohm-m

Eaton's Method: Fracture Pressure

$$p_{ff} = \left(\frac{\mu}{1-\mu}\right)\left(\sigma_{ob} - p_p\right) + p_p$$

p_{ff} Fracture pressure, psi

σ_{ob} Overburden pressure, psi

p_p Pore pressure, psi

μ Poisson's ratio

Horizontal Stress and Pore Pressure

$$\Delta S_{H\max} = \Delta S_{H\min} = \alpha\left(1-2v\right)/\left(1-v\right)\Delta P_p$$

$\Delta S_{H\max}$ Change in greatest horizontal stress, psi

$\Delta S_{H\min}$ Change in least horizontal stress, psi

α Biot poroelastic coefficient $= 1 - K_{\mathrm{dry}}/K_{\mathrm{grain}}$

v Poisson's ratio

ΔP_p Change in pore pressure, psi

K_{dry} Bulk modulus of the dry frame of a porous rock, GPa

K_{grain} Bulk modulus of the grains that make up the rock, GPa

2.3 Drilling Fluids

Mud Rheology

Newtonian Fluid

$$\tau = \mu\gamma$$

Bingham Plastic Model

$$\tau = \tau_y + \mu_p\gamma$$

$$\mu_p = \theta_{600} - \theta_{300}$$

$$\tau_y = \theta_{300} - \mu_p$$

Power-Law Model

$$\tau = K\gamma^n$$

$$n = 3.32\log\frac{\theta_{600}}{\theta_{300}}$$

$$K = \frac{510\times\theta_{300}}{511^n}$$

τ Shear stress, dynes/cm^2, lb/100ft^2

γ Shear rate, sec^{-1}

μ Newtonian viscosity, poise, centipoise

τ_y Yield stress, dynes/cm^2, lb/100ft^2

μ_p Bingham plastic viscosity, poise, centipoise

K Consistency index, lb-sec^n/ft^2

n Flow behavior index, lb-sec^n/ft^2

N Fann rpm

θ_N Fann dial reading at N rpm

Drilling Wastes

$$S = \varepsilon \; HV \,/\, F_s$$

S	Volume of wet drilled solids, bbl
ε	Efficiency of solids control, fraction
HV	Hole volume, bbl
F_s	Fraction of solids in the discard stream

$$L = HV \times (1 - \varepsilon)\,/\,T \qquad \text{or} \qquad L = DS_L \,/\, T$$

L	Liquid discard, bbl
HV	Hole volume, bbl
ε	Efficiency of solids control, fraction
T	Tolerance of the fluid system to solids contamination, fraction
DS_L	Solids buildup in the drilling fluid

Total Solids Generated

$$W_{cg} = 350 Ch \times L \; (1 - \phi) \; SG$$

W_{cg}	Solids generated, lb
Ch	Capacity of hole, bbl/ft
L	Footage drilled, ft
SG	Specific gravity of cuttings
ϕ	Porosity, unitless

Equivalent Circulating Density (ECD)

$$\text{ECD, ppg} = \frac{\text{annular pressure loss, psi}}{0.052 \times \text{True Vertical Depth, ft}} + \text{Mud Weight in Use, ppg}$$

Maximum Allowable Mud Weight From Leak-off Test Data

$$\text{Maximum allowable mud weight, ppg} = \frac{\text{leak-off pressure, psi}}{0.052 \times \text{Casing Shoe TVD, ft}} + \text{Mud Weight, ppg}$$

Hydrostatic Pressure (HP) Decrease When Pulling Pipe Out of the Hole

When Pulling Dry Pipe

HP decrease, psi =

$$\frac{\text{barrels displaced}}{\left(\text{casing capacity, bbl/ft} - \text{pipe displacement, bbl / ft}\right)} \times 0.052 \times \text{mud weight, ppg}$$

barrels displaced =

number of stands pulled $\times$ average length per stand, ft $\times$ pipe displacement, bbl/ft

When Pulling Wet Pipe

$$\text{HP decrease, psi} = \frac{\text{barrels displaced}}{\left(\begin{array}{c}\text{casing} \\ \text{capacity,} \\ \text{bbl/ft}\end{array}\right) - \left(\begin{array}{c}\text{pipe disp., bbl/ft} \\ + \\ \text{pipe cap., bbl/ft}\end{array}\right)} \times 0.052 \times \text{mud weight, ppg}$$

$$\text{barrels displaced} = \begin{matrix} \text{number} \\ \text{of stands} \\ \text{pulled} \end{matrix} \times \begin{matrix} \text{average} \\ \text{length per} \\ \text{stand, ft} \end{matrix} \times \left(\begin{matrix} \text{pipe disp., bbl/ft} \\ + \\ \text{pipe cap., bbl/ft} \end{matrix} \right)$$

Loss of Overbalance Resulting From Falling Mud Level

Feet of Pipe Pulled Dry to Lost Overbalance

$$= \frac{\text{overbalance, psi (casing cap.} - \text{pipe disp., bbl/ft)}}{0.052 \times \text{mud weight, ppg} \times \text{pipe disp., bbl / ft}}$$

Feet of Pipe Pulled Wet to Lost Overbalance

$$= \frac{\text{overbalance, psi (casing cap.} - \text{pipe cap.} - \text{pipe disp., bbl/ft)}}{0.052 \times \text{mud weight, ppg} \times \left(\text{pipe disp.} + \text{pipe cap., bbl / ft}\right)}$$

Formation Temperature (FT)

$$\text{FT, °F} = \left(\begin{matrix} \text{ambient} \\ \text{surface} \\ \text{temperature, °F} \end{matrix} \right) + \left(\begin{matrix} \text{temperature} \\ \text{increase °F per ft of depth} \times \text{TVD, ft} \end{matrix} \right)$$

Increase Mud Density

Mud Weight Increase With 100-lbm Sacks of Barite

$$\text{Barite, sk/100 bbl} = \frac{1,470 \left(\rho_{m2} - \rho_{m1}\right)}{35 - \rho_{m2}}$$

Mud Weight Increase With 100-lbm Sacks of Calcium Carbonate

$$\text{Calcium Carbonate, sk/100 bbl} = \frac{945 \left(\rho_{m2} - \rho_{m1}\right)}{22.5 - \rho_{m2}}$$

ρ_{m2} New fluid weight, ppg
ρ_{m1} Original fluid weight, ppg

Dilution

Mud Weight Reduction With Liquid

$$\text{Water, bbl} = \frac{V_1 \left(W_1 - W_2\right)}{W_2 - D_L}$$

D_L Dilution liquid, ppg
V_1 First mud volume, bbl
W_1 First mud weight, ppg
W_2 Second mud weight, ppg

Mixing Fluids of Different Densities

$$\left(V_1 D_1\right) + \left(V_2 D_2\right) = V_F D_F$$

V_1 Volume of fluid 1 (bbl and gal, for example)
D_1 Density of fluid 1 (bbl and gal, for example)

V_2 Volume of fluid 2 (bbl and gal, for example)

D_2 Density of fluid 1 (bbl and gal, for example)

V_F Volume of final fluid mix

D_F Density of final fluid mix

Will Benefit from UBD	Will Not Benefit from UBD
Formations that usually suffer major formation damage during drilling or completion operations. Wells with skin factors of 5 or higher.	Wells in areas of very low conventional drilling cost.
Formations that exhibit differential sticking tendencies.	Wells drilled in areas of extremely high ROP (that is, ROP≥1,000 ft/day).
Formations with zones with severe losses or fluid invasion from drilling or completion operations.	Extremely high-permeability wells.
Wells with large macroscopic fractures.	Ultralow-permeability wells.
Low-permeability wells.	Poorly consolidated formations.
Wells with massive heterogeneous or highly laminated formations characterized by differing permeabilities, porosities, and pore throat throughput.	Wells with low borehole stability.
High-production reservoirs with low to medium permeabilities.	Wells with loosely cemented laminar boundaries.
Formation, with rock fluid sensitivities.	Wells that contain multiple zones with different pressure regimes.
Formations that exhibit low ROP with OBD.	Reservoirs with interbedded shales or claystones.

Table 2.1—UBD effects for reservoir types (from Lake 2006, Table 12.4)

2.4 Fluid Mechanics

Bit Hydraulics

Total pressure loss excluding bit-nozzle loss (P_c), psi, for a given flow rate Q (gpm)

P_c, psi = circulating pressure, psi – bit nozzle pressure loss, psi

$$M = \frac{\log\left(\dfrac{P_{c1}}{P_{c2}}\right)}{\log\left(\dfrac{Q_1}{Q_2}\right)}$$

Optimum Pressure Losses (P_{opt}), psi, for a Given Maximum Surface Pressure $\left(P_{max}\right)$, psi

For Impact Force

$$P_{opt} = P_{max}\left(\frac{2}{M+2}\right)$$

For Hydraulic Horsepower

$$P_{opt} = P_{max}\left(\frac{1}{M+1}\right)$$

Optimum Flow Rate (Q_{opt}), gpm, for Either Impact Force or Hydraulic Horsepower

$$Q_{opt} = Q_1 \times \left(\frac{P_{opt}}{P_{max}}\right)^{1/M}$$

Pressure at the Bit (P_b), psi

$$P_b = P_{max} - P_{opt}$$

Nozzle Size in 32nd inches for a Given Number N_n and Area A_n, in.^{2} of Nozzles

$$\text{Nozzle Size} = 32\sqrt{\frac{A_n}{0.7854 \times N_n}}$$

Pressure Drop at Bit

$$\Delta P_b = \frac{\rho_m Q^2}{10,858 \times A_n^2}$$

ΔP_b Pressure drop across the bit, psi
ρ_m Mud density, ppg
Q Flow rate, gpm
A_n Nozzle flow area, in.^{2}

Hydraulic Impact Force F_{imp}, lbf

$$F_{\text{imp}} = \frac{Q \times \sqrt{\rho_m \times \Delta P_b}}{57.66}$$

F_{imp} Hydraulic (jet) impact force, lbf
Q Flow rate, gpm
ρ_m Mud density, ppg
ΔP_b Pressure drop across the bit, psi

Pump Calculations

Hydraulic Horsepower (HHP)

$$\text{HHP} = \frac{P \times Q}{1714}$$

P Circulating pressure, psi
Q Circulating rate, gpm

Triplex Pump Output (PO)

Formula 1

$$\text{P.O. (bbl/stk)} = 0.000243 \times D^2 \times S \times E$$

Formula 2

$$\text{P.O. (gpm)} = 0.01010199 \times D^2 \times S \times E$$

D Liner diameter, in.
E efficiency
S Stroke length, in.
SPM   Strokes per minute

Annular Velocity Using Pump Output

$$\text{A.V. (ft/min)} = \frac{24.5 \times Q}{Dh^2 - Dp^2}$$

Q Circulation rate, gpm

Dh   Inside diameter of casing or hole size, in.

Dp   Outsize diameter of pipe, tubing, or collars, in.

$$\frac{dp_f}{dL} = \frac{\tau_g}{300(d_2 - d_1)} \text{ (annulus)},$$

$\dfrac{dp_f}{dL}$ pressure gradient

τ_g gel strength (pounds per 100 square feet)

d_2 internal diameter of casing

d_1 external diameter of drillpipe

2.5 Well Control

Formation Pressure (FP)

FP = Hydrostatic Pressure (HP) + Shut-In DrillPipe Pressure (SIDPP)

Equivalent Circulating Density (ECD)

ECD, ppg = [(Annular Pressure Loss, psi) ÷ 0.052 ÷ TVD, ft] + Current Mud Weight, ppg

Leak-off Test, Equivalent Mud Weight (LOT)

LOT, ppg = (LOT pressure, psi) ÷ 0.052 ÷ (Casing Shoe TVD, ft) + (Mud Weight used for LOT, ppg)

Maximum Initial Shut-in Casing Pressure (MISICP)

This calculation is based on shoe fracture, which is equal to the leak-off test pressure

MISICP, psi = (LOT, ppg – Current Mud Weight, ppg) × 0.052 × Shoe TVD, ft

Slow Circulation Rate (SCR)

SCR, psi = Initial Circulating Pressure, psi – SIDPP, psi

Height of Influx

$$\text{Height of Influx, ft} = \frac{\text{Influx Volume, bbl}}{\text{Annular Capacity Factor, bbl/ft}}$$

Approximate Gas-Migration Rate

$$\text{Approximate gas migration rate, ft/hr} = \frac{\text{Change in Shut-In Casing Pressure, psi}}{\text{Mud Weight, ppg} \times 0.052 \times \text{Time for change, hr}}$$

New Pump Pressure With New Pump Strokes

$$\text{New Pressure, psi} = \text{Original Pressure, psi} \times \left(\frac{\text{New Rate, spm}}{\text{Old Rate, spm}}\right)^2$$

Kill Sheet Calculations

$$KWM = \frac{SIDPP}{(0.052 \times TVD)} + OMW$$

$$ICP = SIDPP + SCR$$

$$FCP = \frac{SCR \times KMW}{OMW}$$

KWM   Kill weight mud, ppg
OMW   Original mud weight, ppg
SIDPP   Shut-in drillpipe pressure, psi
TVD   True vertical depth, ft
ICP   Initial circulating pressure, psi
FCP   Final circulating pressure, psi
SCR   Slow circulating-rate pressure, psi

Bottle Capacity Required

$$\text{Bottle Volume, gal} = \frac{\text{Volume Fluid Required, gal}}{\left(\dfrac{\text{Precharge Pressure, psi}}{\text{Minimum Operating Pressure, psi}}\right) - \left(\dfrac{\text{Precharge Pressure, psi}}{\text{Maximum Operating Pressure, psi}}\right)}$$

Volume of Usable Fluid

Volume of usable fluid, gal

$= \text{Bottle volume, gal} \times$

$$\left(\frac{\text{Precharge Pressure, psi}}{\text{Minimum Operating Pressure, psi}}\right) - \left(\frac{\text{Precharge Pressure, psi}}{\text{Maximum Operating Pressure, psi}}\right)$$

Snubbing Force for Snubbing Operation

Snubbing Force (SF) =

Force at Wellhead (F_p) + Frictional Force − Buoyed Weight of String (W_b)

Force at Wellhead (F_p, lbf)

$$F_p = \frac{\pi \times (\text{OD, in.})^2}{4} \times \text{Wellhead pressure, psi}$$

or

$$F_p = (\text{Pipe OD, in.})^2 \times 0.7854 \times \text{Wellhead pressure, psi}$$

Buoyed Weight W_b of Open-Ended Tubular

$$W_b, \text{lb/ft} = W_{\text{air}}, \text{lb/ft} \times \frac{65.5 - \text{MW, ppg}}{65.5}$$

W_{air} Weight of tubulars in air, lbm/ft
MW   Mud weight, lbm/gal or ppg

Buoyed Weight (W_b) Weight of Close-Ended Tubular Without Fluid in the Pipe

$$W_b, \text{lb/ft} = W_{\text{air}}, \text{lb/ft} - \frac{(\text{Pipe OD, in.})^2}{24.5} \times \text{MW, ppg}$$

Buoyed Weight W_b of Close-Ended Tubular After Filling the Pipe

$$W_b, \text{lb/ft} = W_{\text{air}}, \text{lb/ft} - \frac{(\text{OD, in.})^2 \times \text{Fluid Weight, ppg}}{24.5} - \frac{(\text{ID, in.})^2 \times \text{Fluid Weight, ppg}}{24.5}$$

Balance Point for Close-Ended and Unfilled Pipe

This is where the weight of pipe in the fluid equates to the force created by wellhead pressure.

49

$$L, \text{ft} = \frac{F_p, \text{lb}}{W_{air}, \text{lb/ft} - \left[(OD, \text{in.})^2 \times \text{Fluid Weight}_{well}, \text{ppg} / 24.5\right]}$$

L   Length of drillstring, ft

Balance Point for Close-Ended and Filled Pipe

This is where the weight of pipe in the fluid equates to the force created by wellhead pressure.

$L, \text{ft} =$

$$\frac{F_p, \text{lb}}{W_{air}, \text{lb/ft} - \left[(OD, \text{in.})^2 \times \text{Fluid Weight}_{well}, \text{ppg} / 24.5 - (ID, \text{in.})^2 \times \text{Fluid Weight}_{pipe}, \text{ppg} / 24.5\right]}$$

Maximum Down Force on Jacks

Maximum Down Force, lb $= 0.7 \times$ Critical Buckling Load, lb from pipe data

Effective Area of Snubbing Jacks

Effective Area, sq. in. $=$ Number of jacks $\times 0.7854 \times \left[(ID \text{ Cylinder, in.})^2 - (OD \text{ Rod, in.})^2\right]$

Hydraulic Pressure to Snub

Hydraulic Pressure to Snub, psi $= \dfrac{\text{Maximum Down Force, lb}}{\text{Effective Area, in.}}$

Bullheading Calculations

$$KWM_{pcf} = \frac{\text{Formation Pressure}_{psi}}{0.007 \times \text{Perfs } TVD_{ft}}$$

$$FIT_{psi} = 0.007 \times \left(FIT \text{ } EMW_{pcf} \text{ at perf}\right) \times \text{Perfs } TVD_{ft}$$

$$HP_{psi} = FP_{psi} - SIDPP_{psi}$$

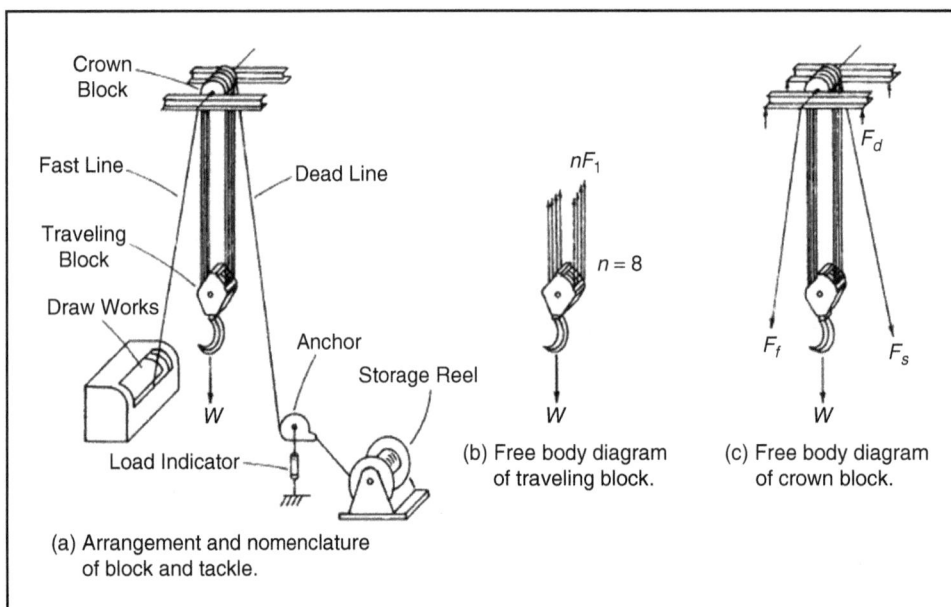

Fig. 2.2—Schematic of block and tackle (Bourgoyne et al. 1986b).

Number of Lines, n	Efficiency, E
6	0.874
8	0.841
10	0.81
12	0.77
14	0.74

Table 2.2—Average efficiency factors for block and tackle systems.

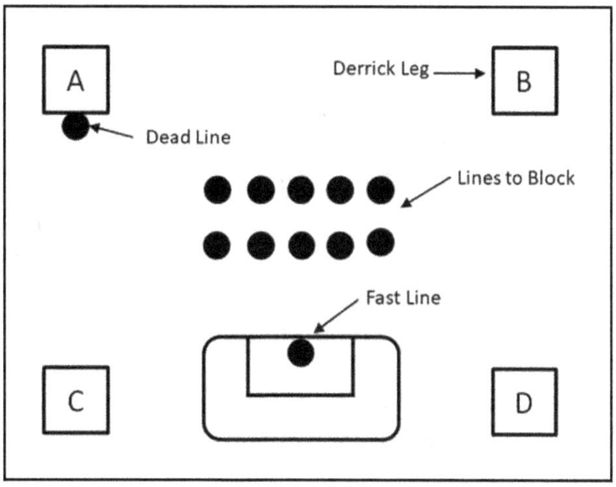

Fig. 2.3—Projection of drilling lines on rig floor (Bourgoyne et al.1986c).

Diameter (in.)	Weight (lbm/ft)	Improved Plow Steel-Breaking Strength (lbf)	Extra-Improved Plow Steel-Breaking Strength (lbf)
½	0.46	23,000	26,600
9/16	0.59	29,000	33,600
⅝	0.72	35,800	41,200
¾	1.04	51,200	58,800
⅞	1.42	69,200	79,600
1	1.85	89,800	103,400
1 ⅛	2.34	113,000	130,000
1 ¼	2.89	138,800	159,800
1 ⅜	3.5	167,000	192,000
1 ½	4.16	197,800	228,000
1 ⅝	4.88	230,000	264,000
1 ¾	5.67	266,000	306,000
1 ⅞	6.5	304,000	348,000
2	7.39	344,000	396,000

Table 2.3—Nominal breaking strength of 6×19* classification wire rope, bright (uncoated) or drawn galvanized, independent wire-rope core.

$$IMDPP_{psi} = FIT_{psi} - HP_{psi}$$

$$KMHP_{psi} = 0.007 \times KWM_{pcf} \times \text{Perfs } TVD_{ft}$$

$$FMDPP_{psi} = FIT_{psi} - KMHP_{psi}$$

KWM	Kill weight mud, lbm/ft^3
FIT	Formation integrity test, psi
EMW	Equivalent mud weight, lbm/ft^3

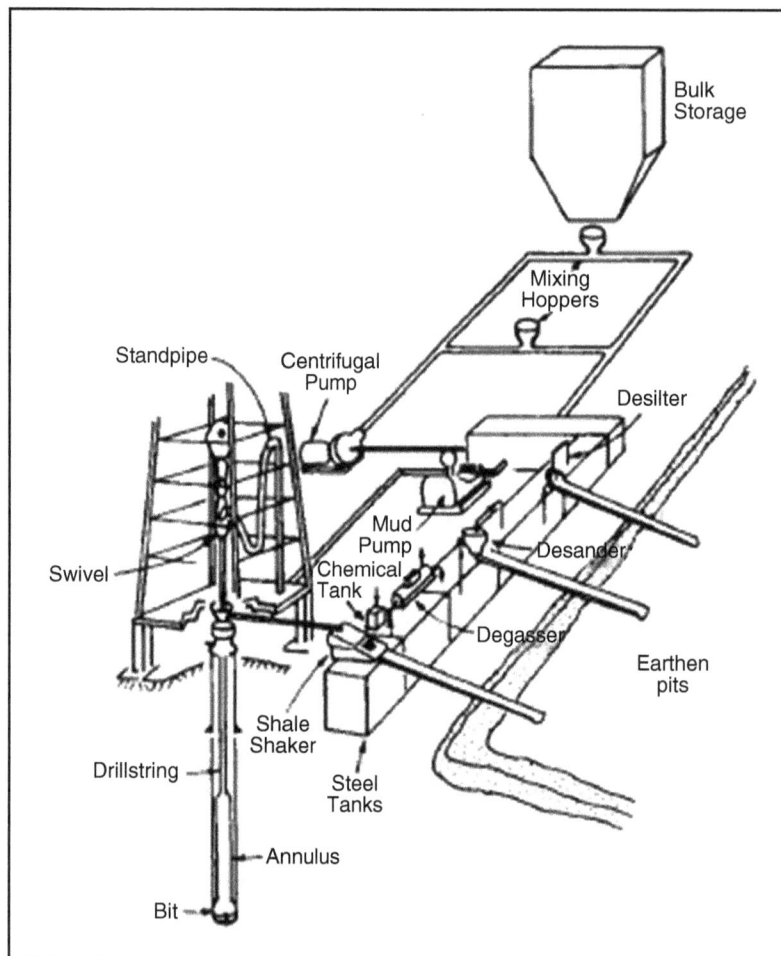

Fig. 2.4—Schematic of example rig-circulating system for liquid drilling fluids (Bourgoyne et al. 1986d).

TVD	True vertical depth, ft
HP	Hydrostatic pressure from *KWM*, psi
FP	Hydrostatic pressure from *KWM*, psi
SIDPP	Shut-in drillpipe pressure, psi
IMDPP	Initial maximum drillpipe pressure, psi
KMHP	Hydrostatic pressure from KWM, psi
FMDPP	Final maximum drillpipe pressure, psi

Lubricate and Bleed Calculations

Cycle Hydrostatic Pressure Gain Per Barrel

$$\Delta HPL_{psi/bbl} = \frac{\text{Gradient Lube Mud}_{psi/ft}}{\text{Annulus Cap}_{bbl/ft} \text{ at top of hole}}$$

Cycle Hydrostatic Pressure Increase or Lubricated Volume To Be Bled Off

$$\Delta HPI_{psi} = \Delta HP_{psi} \times \text{Lubricated Vol}_{bbl}$$

ΔHPL	Hydrostatic pressure loss, psi
ΔHPI	Hydrostatic pressure increase, psi
ΔHP	Hydrostatic pressure gain, psi

2.6 Drilling Mechanics

Rig Equipment

Blocks and Drilling Line

$$M = \frac{W}{F_f}$$

$$F_d = W + F_f + F_s$$

$$F_d = W + \frac{W}{En} + \frac{W}{n} = \left(\frac{1 + E + En}{En}\right)W$$

F_d Load applied to the derrick
W Hook load
F_f Fast line tension
F_s Dead line tension
E Power efficiency

Ton-Mile Calculations

$$T_f = \frac{L_h\left(L_s + L_h\right)W_{dp} + 4L_h\left(W_b + \frac{1}{2}W_1 + \frac{1}{2}W_2 + \frac{1}{2}W_3\right)}{10,560,000}$$

T_f Round-trip ton-miles, ton-miles
L_h Measured depth of the hole or trip depth, ft
L_s Length of the stand, ft
W_{dp} Buoyed weight of the drillpipe per foot, ppf
W_b Weight of block, hook, and others, lbm
W_1 Excess weight of drill collar in mud, lbm
W_2 Excess weight of heavy weight pipe in mud, lbm
W_3 Excess weight of miscellaneous drilling tools in mud, lbm

Drilling or Ton-Miles Calculations

When a hole is drilled only once without any reaming

$$T_d = 2\left(T_{i+1} - T_i\right)$$

When a hole is drilled with one-time reaming

$$T_d = 3\left(T_{i+1} - T_i\right)$$

When a hole is drilled with two-times reaming

$$T_d = 4\left(T_{i+1} - T_i\right)$$

T_d Drilling ton-miles, ton-miles
T_i Round trip calculated, ton-miles

Casing Ton-Miles Calculations

$$T_c = \frac{1}{2}\left[\frac{L_h\left(L_s + L_h\right)W_c + 4L_hW_b}{10,560,000}\right]$$

T_c Casing ton-miles, ton-miles

L_h Measured depth of the hole or trip depth, ft

L_s Length of the stand, ft

W_c Buoyed weight of the casing per foot, ppf

W_b Weight of block, hook, and others, lbm

Drillstring Design

Drill Collar Length

$$L_{dc} = \frac{WOB \times DF}{\omega_{dc} \times BF \times \cos \alpha}$$

L_{dc} Length of the drill collar, ft

WOB Weight on bit, lbm

DF Design factor

ω_{dc} Unit weight of the collar in lbf/ft

BF Buoyancy factor

α Wellbore inclination, degrees

Feet of Drillpipe That Can Be Used With a Specific Bottomhole Assembly (BHA)

$$Length_{max}, ft = \frac{\left[(T \times f) - MOP - W_{bha} \right] \times BF}{W_{dp}}$$

T Tensile strength, lbf for new pipe

f Safety factor to correct new pipe to no. 2 pipe

MOP Margin of overpull

W_{bha} BHA weight in air, lbm/ft

W_{dp} Drillpipe weight in air, lbm/ft, including tool joint

BF Buoyancy factor

Stuck Pipe Calculations

Method 1

$$Free\ pipe, ft = \frac{stretch, in. \times free\ point\ constant}{pull\ force\ in\ thousands\ of\ pounds}$$

$$Free\ point\ constant = A_S \times 2{,}500$$

A_S Pipe wall cross-sectional area, in.^{2}

Method 2

$$Free\ pipe, ft = \frac{735294 \times e \times W_{dp}}{differential\ pull, lb}$$

e Pipe stretch, in.

W_{dp} Drillpipe weight, lbm/ft (plain end)

2.7 Tubular Mechanics

Tension

$$F_e = \sum \left(W_s \cos \alpha + F_D + DF_{area} \right) - F_{bottom} - WOB + F_{bs}$$

F_e Effective tension, lbf

W_s Air weight of the segment, $= L\omega_{air}$

L	Length of drillstring hanging below point, ft
ω_{air}	Weight per foot of drillstring in air, lbm/ft
F_D	Drag force, lbf
ΔF_{area}	Change in force resulting from the change in area, lbf
F_{bottom}	Bottom pressure force, lbf
WOB	Weight on bit, lbm
F_{bs}	Buckling stability force, lbf
α	Well inclination, degrees

Slip Crushing

$$F_{max} = \frac{F_y}{\sqrt{1 + \frac{2D_p^2 fA_p}{\left(D_p^2 - D_i^2\right)A_s} + \left[\frac{2D_p^2 fA_p}{\left(D_p^2 - D_i^2\right)A_s}\right]^2}}$$

F_{max}	Maximum allowable static axial load, lbf
F_y	Tensile strength of the pipe, lbf
A_s	Contact area between slip and pipe, in.^{2}, $= \pi D_p L_s$
A_p	Cross-sectional area of the pipe, in.^{2}
D_p	OD of the pipe, in.
D_i	ID of the pipe, in.
L_s	Length of the slips, in.
f	Lateral load factor of slips, $= \left(1 - \mu \tan a\right)/\left(\mu + \tan a\right)$
μ	Coefficient of friction between slips and bushings
a	Slip taper angle, degrees

2.8 Casing Design

Tension

Axial Force

$$F_{ten} = \sigma_{yield} A_s$$

F_{ten}	Axial force, lbf
σ_{yield}	Minimum yield strength, psi
A_s	Pipe cross-sectional area, in.^{2}

Suspended Weight

$$F_a = F_{air} - F_{bu}$$

F_a	Resultant axial force, lbf
F_{air}	Weight of the string in the air, lbf
F_{bu}	Buoyancy force, lbf

von Mises Stress

$$\sigma_{vm} = \frac{1}{\sqrt{2}} \sqrt{\left(\sigma_t - \sigma_a\right)^2 + \left(\sigma_r - \sigma_t\right)^2 + \left(\sigma_a - \sigma_r\right)^2}$$

σ_{vm}	von Mises (effect combined) stress, psi
σ_r	Radial stress, psi
σ_t	Tangential stress, psi
σ_a	Axial stress, psi

			Empirical Coefficients		
Grade	F_1	F_2	F_3	F_4	F_5
J-55	2.991	0.0541	1,206	1.989	0.036
L-80	3.071	0.0667	1,955	1.998	0.0434
C-95	3.124	0.0743	2,404	2.029	0.0482
P-110	3.181	0.0819	2,852	2.066	0.0532

Table 2.4—Empirical coefficients used for collapse-pressure determination.

Grade	Yield	Plastic Transition	Elastic
J-55	14.81	25.01	37.21
L-80	13.38	22.47	31.02
C-95	12.85	21.33	28.36
P-110	12.44	20.41	26.22

Table 2.5—Range of d_n/t for various collapse-pressure regions when axial stress is zero.

Bending Force

Maximum Bending Stress

$$F_{ab} = 218\alpha d_n A_s$$

F_{ab} Equivalent axial force, lbf
α Dogleg severity, deg/100 ft
d_n OD, in.
A_s Cross-sectional area, in.^{2}

Burst Pressure

$$P_{br} = 0.875 \frac{2\sigma_{yield}t}{(d_n)}$$

P_{br} Burst pressure rating as defined by the American Petroleum Institute (API), psi
d_n OD, in.
t Cylinder wall thickness, in.
σ_{yield} Minimum yield strength, psi

Collapse Pressure

Effective Yield Strength in the Presence of Significant Axial Stress

$$(\sigma_{yield})_e = \sigma_{yield} \left\{ \left[1 - 0.75 \left(\frac{\sigma_a}{\sigma_{yield}} \right)^2 \right]^{0.5} - 0.5 \left(\frac{\sigma_a}{\sigma_{yield}} \right) \right\}$$

Elastic Collapse

$$p_{cr} = \frac{46.95 \times 10^6}{\left(\dfrac{d_n}{t} \right) \left(\dfrac{d_n}{t} - 1 \right)^2}$$

Transition Collapse

$$p_{cr} = (\sigma_{yield})_e \left(\frac{F_4}{d_n/t} - F_5 \right)$$

56

Plastic Collapse

$$p_{cr} = \left(\sigma_{yield}\right)_e \left(\frac{F_1}{d_n / t} - F_2\right) - F_3$$

Yield Collapse

$$p_{cr} = 2\left(\sigma_{yield}\right)_e \left[\frac{d_n / t - 1}{\left(d_n / t\right)^2}\right]$$

p_{cr}	Collapse pressure rating, psi
d_n	OD, in.
t	Wall thickness, in.
σ_{yield}	Minimum yield strength, psi
$\left(\sigma_{yield}\right)_e$	Effective yield strength in the presence of significant axial stress, psi
σ_a	Axial stress, psi
$F_1 ... F_5$	Empirical coefficient from tables using grade and d_o / t

Special Design Consideration

Changing Internal Pressure

$$\Delta F_a = p_i \pi d^2 / 4$$

ΔF_a	Change in axial force, lbf
d	ID of the pipe, in.
p_i	Change in internal pressure, psi

2.9 Cementing

Cement Slurry Requirements

Slurry Density

$$\rho_{sl} = \frac{W_c + W_w + W_a}{V_c + V_w + V_a}$$

ρ_{sl}	Slurry density, ppg
W_c	Weight of cement, lbm
W_w	Weight of water, lbm
W_a	Weight of additive, lbm
V_c	Cement volume, gal
V_w	Water volume, gal
V_a	Additive volume, gal

Contact Time

$$V_t = t_c \times q \times 5.615$$

V_t	Volume of cement needed for removal of mudcake by turbulent flow, ft³
t_c	Contact time, min
q	Displacement rate, bbl/min

Cementing Calculations

Cement Additives

Weight of additives per sack of cement, lb = percent of additive 94 lb/sk

$$\text{Total water requirement gal/sk, of cement} = \frac{\text{Cement water requirement, gal/sk}}{} + \frac{\text{Additive water requirement, gal/sk}}{}$$

$$\frac{\text{Volume of slurry, gal/sk}}{} = \frac{94\ \text{lb}}{\text{SG of cement} \times 8.33\ \text{lb/gal}} + \frac{\text{weight od additive, lb}}{\text{SG of additive} \times 8.33\ \text{lb/gal}} + \text{water volume, gal}$$

$$\text{Slurry yield, ft}^3/\text{sk} = \frac{\text{vol of slurry, gal/sk}}{7.48\ \text{gal/ft}^3}$$

$$\text{Slurry density, lb/gal} = \frac{94 + \text{wt of additive} + (8.33 \times \text{vol of water/sk})}{\text{vol of slurry, gal/sk}}$$

Water Requirements

Weight of materials, lb/sk = 94 + (8.33 × vol of water, gal) + (% of additive 94)

$$\text{Volume of slurry, gal/sk} = \frac{94\ \text{lb/sk}}{\text{SG} \times 8.33} + \frac{\text{wt of additive, lb/sk}}{\text{SG} \times 8.33} + \text{water vol, gal}$$

Balanced Plug and Formula

Height of Cement

$$h = \frac{N}{C+T}$$

h Height of cement, ft
C Annular capacity, ft³/ft
T Tubing capacity, ft³/ft
N Volume of cement in ft³ or sacks × yield

2.10 Well Planning

Well Path Design

q = build rate in °/100 ft

$$r_1 = \frac{180}{\pi} \times \frac{1}{q}$$

$$L_{DC} = \frac{\pi}{180} \times r_1 \times \theta$$

or

$$L_{DC} = \frac{\theta}{q}$$

$$\tan \Omega = \frac{CO}{L_{CB}} = \frac{r_1}{L_{CB}}$$

Bentonite	Maximum Water Requirements		Slurry Weight		Slurry Volume
%	gal/sack	ft³/sack	lbm/gal	lbm/ft³	ft³/sack
0	5.2	0.70	15.6	117	1.18
2	6.5	0.87	14.7	110	1.36
4	7.8	1.04	14.1	105	1.55
6	9.1	1.22	13.5	101	1.73
8	10.4	1.39	13.1	98	1.92

Table 2.6a—Class A with bentonite.

Bentonite	API Casing Tests		
%	4,000 ft	6,000 ft	8,000 ft
0	3:00	2:25	1:40
2	2:25	1:48	1:34
4	2:34	1:57	1:32
6	2:35	1:45	1:22
8	2:44	1:50	1:24

Table 2.6b—Class A thickening time—hr:min (pressure-temperature-thickening-time test).

Bentonite	Maximum Water Requirements		Slurry Weight		Slurry Volume
%	gal/sack	ft³/sack	lbm/gal	lbm/ft³	ft³/sack
0	4.3	0.58	16.4	123	1.06
2	5.49	0.73	15.5	115	1.22
4	6.69	0.89	14.7	110	1.38
6	7.88	1.05	14.1	105	1.55
8	9.07	1.21	13.6	101	1.73

Table 2.7a—Class H with bentonite.

Bentonite	API Casing Tests		
%	4,000 ft	6,000 ft	8,000 ft
0	3:04	2:14	1;35
2	2:50	2:00	1:20
4	3:05	2:10	1:25
6	3:05	2:15	1:25
8	3:10	2:15	1:30

Table 2.7b—Class H thickening time—hr:min (pressure-temperature-thickening-time test).

$$L_{CB} = \frac{D_3 - D_2}{\cos\theta}$$

$$X_2 = r_1 - r_1\cos\theta = r_1\left(1 - \cos\theta\right)$$

$$D_2 = D_1 + r_1\sin\theta$$

Total measured depth for a TVD of $D_3 = D_m$

$$= D_1 + \theta/q + r_1/\tan\Omega = D_1 + \theta/q + \frac{D_3 - D_2}{\cos\theta}$$

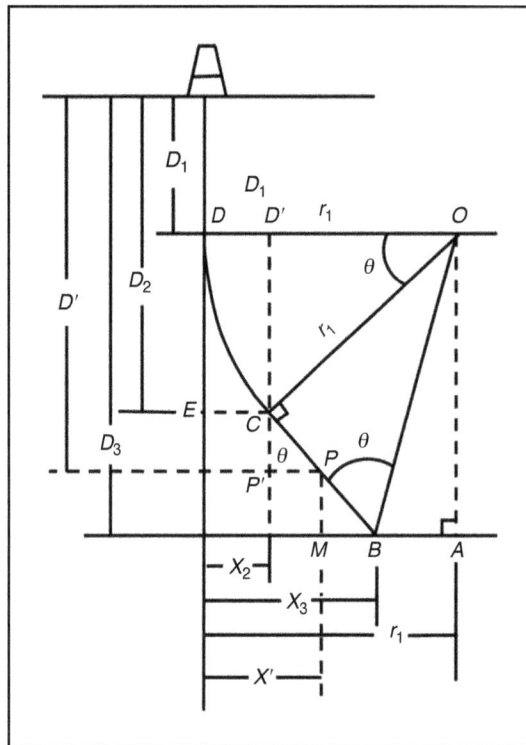

Fig. 2.5—Geometry of build-and-hold-type well path for $X_3 < r_1$.

Total departure at $D_m = X_3$

$$X_3 = X_2 + L_{CB} \sin\theta$$

Calculations before reaching final build angle (Point C′).

q = build rate in °/100 ft

$$D_n = D_1 + r_1 \sin\theta'$$

$$X_N = r_1(1 - \cos\theta')$$

Measured depth at C′

$$D_{MN} = D_1 + \left(\frac{\theta'}{q}\right)$$

Well Cost

Drilling Cost (Cost Per Foot)

$$C_{di} = \frac{C_{bi} + C_r\left(T_{di} + T_{ti} + T_{ci}\right)}{D_i}$$

i	Drill bit number
C_{di}	Drilling cost, USD/ft
C_{bi}	Bit cost, USD

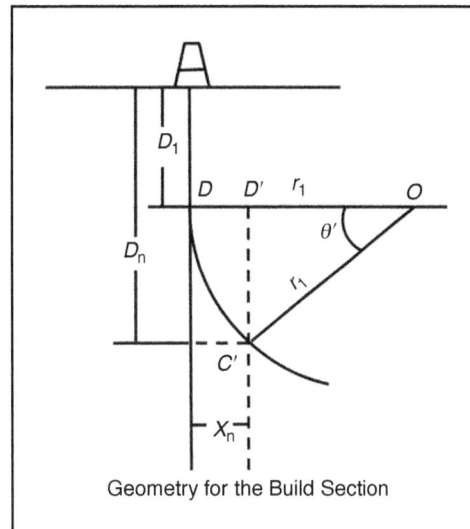

Geometry for the Build Section

Fig. 2.6—Geometry for the build section.

C_r Rig cost, USD/hr
T_{di} Drilling time, hr
T_{ti} Trip time, hr
T_{ci} Connection time, hr
ΔD_i Formation interval drilled, ft

$$T_{ti} = 2\left(\frac{t_s}{L_s}\right) D_i$$

T_{ti} Trip time required to change a bit and resume drilling operations, hr
t_s Average time required to handle one stand of drillstring, hr
L_s Average length of one stand of drillstring, ft
D_i Depth to which the trip was made, ft

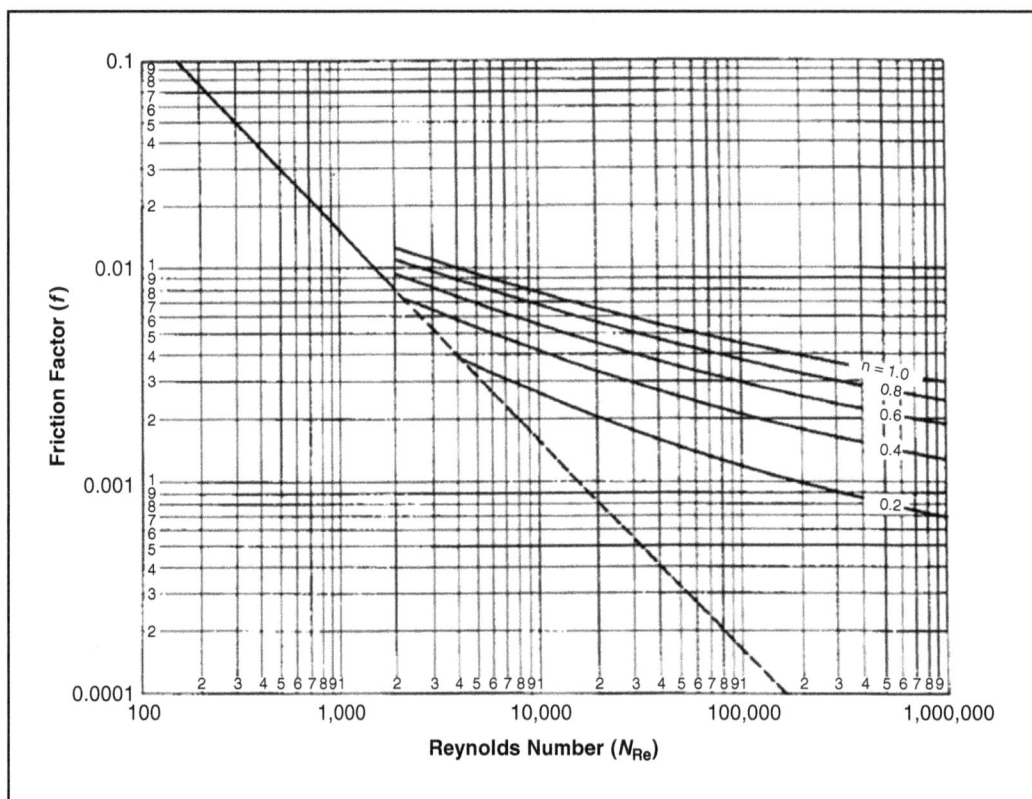

Fig. 2.7—Friction factor vs. Reynolds number (Bourgoyne et al. 1986e).

Nominal size, in.	ID, in.	Wall thickness, in.	Approximate weight per foot, lbm/ft	Capacity, bbl/ft	Displacement, bbl/ft
3.5	2.0625	0.719	25.3	0.0042	0.0092
3.5	2.25	0.625	23.2	0.005	0.0084
4	2.5625	0.719	27.2	0.0073	0.0100
4.2	2.75	0.875	41.0	0.0074	0.0149
5	3	1.000	49.3	0.0088	0.0179
5.5	3.375	1.063	57.0	0.0111	0.0207
6.625	4.5	1.063	70.8	0.0196	0.0257

Table 2.8—Heavy weight drillpipe properties.

OD, in.	Weight, lbm/ft		Collapse, psi — Grade				Internal Yield, psi — Grade				Tensile, kips — Grade			
			D	E	G	S-135	D	E	G	S-135	D	E	G	S-135
2⅜	4.85	1.995	6,850	11,040	13,250	16,560	7,110	10,500	14,700	18,900	70	98	137	176
2⅜	6.65	1.815	11,440	15,600	18,720	23,400	11,350	15,470	21,660	27,850	101	138	194	249
2⅞	6.85	2.441	–	10,470	12,560	15,700	–	9,910	13,870	17,830	–	136	190	245
2⅞	10.4	2.151	12,110	16,510	19,810	24,760	12,120	16,530	23,140	29,750	157	214	300	386
3½	9.5	2.992	–	10,040	12,110	15,140	–	9,520	13,340	17,140	–	194	272	350
3½	13.3	2.764	10,350	14,110	16,940	21,170	10,120	13,800	19,320	24,840	199	272	380	489
3½	15.5	2.602	12,300	16,770	20,130	25,160	12,350	16,840	23,570	30,310	237	323	452	581
4	11.85	3.476	–	8,410	10,310	12,820	–	8,600	12,040	15,470	–	231	323	415
4	14	3.34	8,330	11,350	14,630	17,030	7,940	10,830	15,160	19,500	209	285	400	514
4½	13.75	3.958	–	7,200	8,920	10,910	–	7,900	11,070	14,230	–	270	378	486
4½	16.6	3.826	7,620	10,390	12,470	15,590	7,210	9,830	13,760	17,690	242	331	463	595
4½	20	3.64	9,510	12,960	15,560	19,450	9,200	12,540	17,560	22,580	302	412	577	742
5	16.25	4.408	–	6,970	8,640	10,550	–	7,770	10,880	13,980	–	328	459	591
5	19.5	4.276	7,390	10,000	12,090	15,110	6,970	9,500	13,300	17,100	290	396	554	712
5½	21.9	4.778	6,610	8,440	10,350	12,870	6,320	8,610	12,060	15,500	321	437	612	787
5½	24.7	4.67	7,670	10,460	12,560	15,700	7,260	9,900	13,860	17,820	365	497	696	895

Table 2.9a—Drillpipe properties (Bourgoyne et al. 1986f).

OD, in.	Weight, lbm/ft	ID, in.	Capacity, bbl/ft	Displacement, bbl/ft	Closed-end displacement, bbl/ft
2 ⅜	4.85	1.995	0.00387	0.0016	0.0055
2 ⅜	6.65	1.815	0.0032	0.0025	0.0055
2 ⅞	6.85	2.441	0.00579	0.0022	0.008
2 ⅞	10.40	2.151	0.00449	0.004	0.008
3 ½	9.50	2.992	0.0087	0.0035	0.0119
3 ½	13.30	2.764	0.00742	0.0048	0.0119
3 ½	15.50	2.602	0.00658	0.0056	0.0119
4	11.85	3.476	0.01174	0.0038	0.0155
4	14.00	3.340	0.01084	0.0051	0.0155
4 ½	13.75	3.958	0.01522	0.005	0.0197
4 ½	16.6	3.826	0.01422	0.0064	0.0197
4 ½	20	3.640	0.01287	0.0073	0.0197
5	16.25	4.408	0.01888	0.0054	0.0243
5	19.5	4.276	0.01776	0.0065	0.0243
5 ½	21.9	4.778	0.02218	0.008	0.0294
5 ½	24.7	4.670	0.02119	0.0082	0.0294

Table 2.9b—Drillpipe properties (Bourgoyne et al. 1986f).

Range	Length, ft
1	18 to 20
2	27 to 30
3	38 to 45

Table 2.10—Drillpipe range (Bourgoyne et al. 1986g).

OD	ID Capacity	In bbl / ft	1½ 0.0022	1¾ 0.003	2 0.0039	2¼ 0.0049	2½ 0.0061	2¾ 0.0073	3 0.0087	3¼ 0.0103	3½ 0.0119	3¾ 0.0137
4.00	Weight	lbm/ft	36.7	34.5	32	29.2	–	–	–	–	–	–
	Displacement	bbl/ft	0.0133	0.0125	0.0116	0.0106	–	–	–	–	–	–
	CE Displacement	bbl/ft	0.0155	0.0155	0.0155	0.0155	–	–	–	–	–	–
4.50	Weight	lbm/ft	48.1	45.9	43.4	40.6	–	–	–	–	–	–
	Displacement	bbl/ft	0.0175	0.0167	0.0158	0.0148	–	–	–	–	–	–
	CE Displacement	bbl/ft	0.0197	0.0197	0.0197	0.0197	–	–	–	–	–	–
4.75	Weight	lbm/ft	54.3	52.1	49.6	46.8	43.6	–	–	–	–	–
	Displacement	bbl/ft	0.0197	0.0189	0.0181	0.017	0.0159	–	–	–	–	–
	CE Displacement	bbl/ft	0.0219	0.0219	0.0219	0.0219	0.0219	–	–	–	–	–
5.00	Weight	lbm/ft	60.8	58.6	56.1	53.5	50.1	–	–	–	–	–
	Displacement	bbl/ft	0.0221	0.0213	0.0204	0.0194	0.0182	–	–	–	–	–
	CE Displacement	bbl/ft	0.0243	0.0243	0.0243	0.0243	0.0243	–	–	–	–	–
5.50	Weight	lbm/ft	74.8	72.6	70.1	67.3	64.1	60.6	56.8	–	–	–
	Displacement	bbl/ft	0.0272	0.0264	0.0255	0.0245	0.0233	0.0221	0.0207	–	–	–
	CE Displacement	bbl/ft	0.0294	0.0294	0.0294	0.0294	0.0294	0.0294	0.0294	–	–	–
5.75	Weight	lbm/ft	82.3	80.1	77.6	74.8	71.6	68.1	64.3	–	–	–
	Displacement	bbl/ft	0.0299	0.0291	0.0282	0.0272	0.0261	0.0248	0.0234	–	–	–
	CE Displacement	bbl/ft	0.0321	0.0321	0.0321	0.0321	0.0321	0.0321	0.0321	–	–	–
6.00	Weight	lbm/ft	90.1	87.9	85.4	82.6	79.4	75.9	72.1	67.9	63.4	–
	Displacement	bbl/ft	0.0328	0.032	0.0311	0.0301	0.0289	0.0276	0.0262	0.0247	0.0231	–
	CE Displacement	bbl/ft	0.035	0.035	0.035	0.035	0.035	0.035	0.035	0.035	0.035	–
6.50	Weight	lbm/ft	107	104.8	102.3	99.5	96.3	92.8	89	84.8	80.3	–
	Displacement	bbl/ft	0.0389	0.0381	0.0372	0.0362	0.035	0.0338	0.0324	0.0308	0.0292	–
	CE Displacement	bbl/ft	0.0411	0.0411	0.0411	0.0411	0.0411	0.0411	0.0411	0.0411	0.0411	–
6.75	Weight	lbm/ft	116	113.8	111.3	108.5	105.3	101.8	98	93.8	89.3	–
	Displacement	bbl/ft	0.0422	0.0414	0.0405	0.0395	0.0383	0.037	0.0356	0.0341	0.0325	–
	CE Displacement	bbl/ft	0.0444	0.0444	0.0444	0.0444	0.0444	0.0444	0.0444	0.0444	0.0444	–
7.00	Weight	lbm/ft	125	122.8	120.3	117.5	114.3	110.8	107	102.8	98.3	93.4
	Displacement	bbl/ft	0.0455	0.0447	0.0438	0.0427	0.0416	0.0403	0.0389	0.0374	0.0358	0.034
	CE Displacement	bbl/ft	0.0476	0.0476	0.0476	0.0476	0.0476	0.0476	0.0476	0.0476	0.0476	0.0476
7.50	Weight	lbm/ft	144	141.8	139.3	136.5	133.3	129.8	126	121.8	117.3	112.4
	Displacement	bbl/ft	0.0524	0.0516	0.0507	0.0497	0.0485	0.0472	0.0458	0.0443	0.0427	0.0409
	CE Displacement	bbl/ft	0.0546	0.0546	0.0546	0.0546	0.0546	0.0546	0.0546	0.0546	0.0546	0.0546
8.00	Weight	lbm/ft	165	162.8	160.3	157.5	154.3	150.8	147	142.8	138.3	133.4
	Displacement	bbl/ft	0.06	0.0592	0.0583	0.0573	0.0561	0.0549	0.0535	0.052	0.0503	0.0485
	CE Displacement	bbl/ft	0.0622	0.0622	0.0622	0.0622	0.0622	0.0622	0.0622	0.0622	0.0622	0.0622
8.50	Weight	lbm/ft	187	184.8	182.3	179.5	176.3	172.8	169	164.8	160.3	155.4
	Displacement	bbl/ft	0.068	0.0672	0.0663	0.0653	0.0641	0.0629	0.0615	0.06	0.0583	0.0565
	CE Displacement	bbl/ft	0.0702	0.0702	0.0702	0.0702	0.0702	0.0702	0.0702	0.0702	0.0702	0.0702
9	Weight	lbm/ft	210.2	208	205.6	202.7	199.6	196	192.2	188	183.5	178.7
	Displacement	bbl/ft	0.0765	0.0757	0.0748	0.0738	0.0726	0.0714	0.07	0.0685	0.0668	0.0651
	CE Displacement	bbl/ft	0.0787	0.0787	0.0787	0.0787	0.0787	0.0787	0.0787	0.0787	0.0787	0.0787

Table 2.11—Drill collars capacity and displacement.

API Grade	Yield Strength, kpsi		Minimum Ultimate Tensile Strength, kpsi	Minimum Elongation, %
	Minimum	Maximum		
H-40	40	80	60	29.5
J-55	55	80	75	24.0
K-55	55	80	95	19.5
C-75	75	90	95	19.5
L-80	80	95	95	19.5
N-80	80	110	100	18.5
C-90	90	105	100	18.5
C-95	95	110	105	18.0
P-110	110	140	125	15.0

Table 2.12—Casing properties, API (Bourgoyne et al. 1986h).

OD, in.	Nominal Weight, Tube and Coupling, lbm/ft	Wall Thickness, in.	ID, in.	Drift, in.
5	13.0	0.253	4.494	4.369
	15.0	0.296	4.408	4.283
	18.0	0.362	4.276	4.151
	21.4	0.437	4.126	4.001
	23.2	0.478	4.044	3.919
	24.1	0.500	4.000	3.875
5 ½	15.5	0.275	4.950	4.825
	17.0	0.304	4.892	4.767
	20.0	0.361	4.778	4.653
	23.0	0.415	4.670	4.545
7	23.0	0.317	6.366	6.241
	26.0	0.362	6.276	6.151
	29.0	0.408	6.184	6.059
	32.0	0.453	6.094	5.969
	35.0	0.498	6.004	5.879
	38.0	0.540	5.920	5.795
7 ⅝	26.4	0.328	6.969	6.844
	29.7	0.375	6.875	6.750
	33.7	0.430	6.765	6.640
	39.0	0.500	6.625	6.500
	42.8	0.562	6.501	6.376
	47.1	0.625	6.375	6.250
9 ⅝	36.0	0.352	8.921	8.765
	40.0	0.395	8.835	8.679
	43.5	0.435	8.755	8.599
	47.0	0.472	8.681	8.525
	53.5	0.545	8.535	8.379
10 ¾	55.5	0.495	9.760	9.604
11 ¾	60.0	0.489	10.772	10.616
	65.0	0.534	10.620	10.526
13 ⅜	61.0	0.430	12.515	12.359
	68.0	0.480	12.415	12.259
	72.0	0.514	12.347	12.191
16	75.0	0.438	15.124	14.936
	84.0	0.495	15.010	14.822
20	94.0	0.438	19.124	18.936
	106.5	0.500	19.000	18.812
	133.0	0.635	18.730	18.542

Table 2.13—Casing properties.

OD, in.	Nominal Weight, T&C, lbm/ft	Grade	Collapse Resistance, psi	Pipe-Body Yield Strength, 1,000 lbf	Internal Pressure Resistance, Plain End, psi
5	11.5	J-55	3,060	182	4,240
	13.0	J-55	4,140	208	4,870
	15.0	J-55	5,560	241	5,700
	15.0	L-80	7,250	350	8,290
	18.0	L-80	10,500	422	10,140
	21.4	L-80	12,760	501	12,240
	23.2	L-80	13,830	543	13,380
	24.1	L-80	14,400	566	14,000
	15.0	C-95	8,110	416	9,840
	18.0	C-95	12,030	501	12,040
	21.4	C-95	15,160	595	14,530
	23.2	C-95	16,430	645	15,890
	24.1	C-95	17,100	672	16,630
	15.0	P-110	8,850	481	11,400
	18.0	P-110	13,470	580	13,940
	21.4	P-110	17,550	689	16,820
	23.2	P-110	19,020	747	18,400
	24.1	P-110	19,800	778	19,250
5 ½	15.5	J-55	4,040	248	4,810
	17.0	J-55	4,910	273	5,320
	17.0	L-80	6,280	397	7,740
	20.0	L-80	8,830	466	9,190
	23.0	L-80	11,160	530	10,560
	17.0	C-95	6,940	471	9,190
	20.0	C-95	10,010	554	10,910
	23.0	C-95	12,940	630	12,540
	17.0	P-110	7,480	546	10,640
	20.0	P-110	11,100	641	12,640
	23.0	P-110	14,540	729	14,520

Table 2.14a—Casing properties.

OD, in.	Nominal Weight, T&C, lbm/ft	Grade	Collapse Resistance, psi	Pipe Body Yield Strength, 1,000 lbf	Internal Pressure Resistance, Plain End, psi
7	20	J-55	2,270	316	3,740
	23	J-55	3,270	366	4,360
	26	J-55	4,320	415	4,980
	23	L-80	3,830	532	6,340
	26	L-80	5,410	604	7,240
	29	L-80	7,020	676	8,160
	32	L-80	8,610	745	9,060
	35	L-80	10,810	814	9,960
	38	L-80	11,390	877	10,800
	23	C-95	4,140	632	7,530
	26	C-95	5,880	717	8,600
	29	C-95	7,830	803	9,690
	32	C-95	9,750	885	10,760
	35	C-95	11,650	966	11,830
	38	C-95	13,440	1,041	12,820
	26	P-110	6,230	830	9,960
	29	P-110	8,530	929	11,220
	32	P-110	10,780	1,025	12,460
	35	P-110	13,020	1,119	13,700
	38	P-110	15,140	1,205	14,850
7 5/8	26.4	L-80	3,400	602	6,020
	29.7	L-80	4,790	683	6,890
	33.7	L-80	6,560	778	7,900
	39.0	L-80	8,820	895	9,180
	42.8	L-80	10,810	998	10,320
	45.3	L-80	11,510	1,051	10,920
	47.1	L-80	12,040	1,100	11,480
	26.4	C-95	3,710	714	7,150
	29.7	C-95	5,140	811	8,180
	33.7	C-95	7,280	923	9,380
	39.0	C-95	10,000	1,063	10,900
	42.8	C-95	12,410	1,185	12,250
	45.3	C-95	13,660	1,248	12,970
	47.1	C-95	14,300	1,306	13,630
	29.7	P-110	5,350	940	9,470
	33.7	P-110	7,870	1,069	10,860
	39.0	P-110	11,080	1,231	12,620
	42.8	P-110	13,920	1,372	14,190
	45.3	P-110	15,430	1,446	15,020
	47.1	P-110	16,550	1,512	15,780

Table 2.14b—Casing properties.

OD, in.	Nominal Weight, T&C, lbm/ft	Grade	Collapse Resistance, psi	Pipe Body Yield Strength, 1,000 lbf	Internal Pressure Resistance, Plain End, psi
9 ⅝	36.0	J-55	2,020	564	3,520
	40.0	J-55	2,570	630	3,950
	40.0	L-80	3,090	916	5,750
	43.5	L-80	3,810	1,005	6,330
	47.0	L-80	4,760	1,086	6,870
	53.5	L-80	6,620	1,244	7,930
	40.0	C-95	3,320	1,088	6,820
	43.5	C-95	4,120	1,193	7,510
	47.0	C-95	5,090	1,289	8,150
	53.5	C-95	7,340	1,477	9,410
	43.5	P-110	4,420	1,381	8,700
	47.0	P-110	5,300	1,493	9,440
	53.5	P-110	7,950	1,710	10,900
10 ¾	55.5	L-80	4,020	1,276	6,450
	55.5	C-95	4,290	1,515	7,660
	55.5	P-110	4,610	1,754	8,860
11 ¾	60.0	J-55	2,660	952	4,010
	60.0	L-80	3,180	1,384	5,830
	65.0	L-80	3,870	1,505	6,360
	60.0	C-95	3,440	1,644	6,920
	65.0	C-95	4,180	1,788	7,560
	60.0	P-110	3,610	1,903	8,010
	65.0	P-110	4,480	2,070	8,750
13 ⅜	54.5	J-55	1,130	853	2,730
	61.0	J-55	1,540	962	3,090
	68.0	J-55	1,950	1,069	3,450
	68.0	L-80	2,260	1,556	5,020
	72.0	L-80	2,670	1,661	5,380
	68.0	C-95	2,330	1,847	5,970
	72.0	C-95	2,820	1,973	6,390
	68.0	P-110	2,330	2,139	6,910
	72.0	P-110	2,890	2,284	7,400
16	75.0	J-55	1,020	1,178	2,630
	84.0	J-55	1,410	1,326	2,980
20	94.0	J-55	520	1,480	2,110
	106.5	J-55	770	1,685	2,410
	133.0	J-55	1,500	2,125	3,060

Table 2.14c—Casing properties.

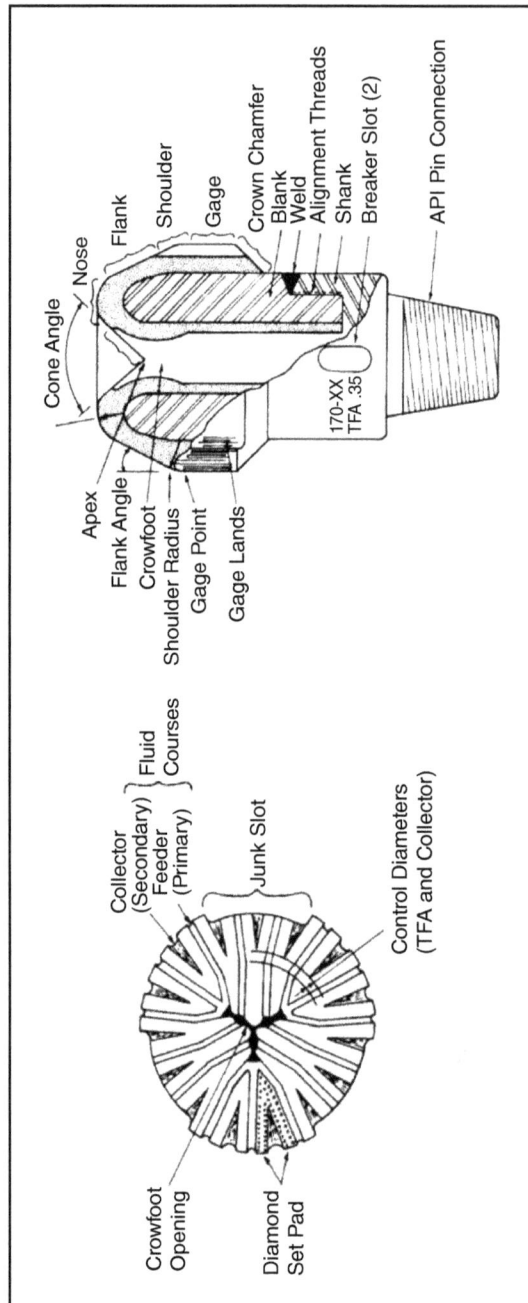

Fig. 2.8—Diamond cutter drag bit, design nomenclature (Bourgoyne et al. 1986i).

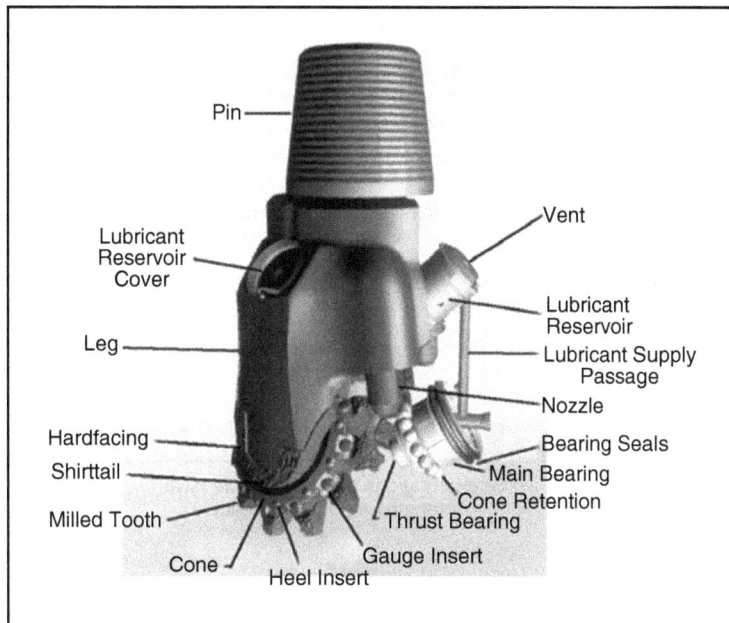

Fig. 2.9—Parts of a roller-cone bit (Lake 2006).

Formation	International Association of Drilling Contractors (IADC) Series Number	
Soft	D1	0
		1
		2
		3
		4
Medium Soft	D2	0
		1
		2
		3
		4
Medium	D3	0
		1
		2
		3
		4
Medium Hard	D4	0
		1
		2
		3
		4
Hard	D5	0
		1
		2
		3
		4

Table 2.15—Diamond drillbit classification chart (courtesy of IADC).

SPE Petroleum Engineering Certification and PE License Exam Reference Guide

	Series	Formations	Types
Milled Tooth Bits	1	Soft formations with low compressive strength and high drillability.	1
			2
			3
			4
	2	Medium to medium-hard formations with high compressive strength.	1
			2
			3
			4
	3	Hard semi-abrasive and abrasive formations.	1
			2
			3
			4
Insert Bits	4	Soft formations with low compressive strength and high drillability.	1
			2
			3
			4
	5	Soft to medium-hard formations with low compressive strength.	1
			2
			3
			4
	6	Medium-hard formations with high compressive strength.	1
			2
			3
			4
	7	Hard semi-abrasive and abrasive formations.	1
			2
			3
			4
	8	Extremely hard and abrasive formations.	1
			2
			3
			4

Table 2.16—Rolling-cutter-bit classification chart (courtesy of IADC).

IADC Dull Grading

Inner Cutting Structure (1) (All Inner Rows)
(For fixed cutter bits, use the inner 2/3 of the bit radius)

Outer Cutting Structure (2) (Gage Row Only)
(For fixed cutter bits, use the outer 1/3 of the bit radius)

In columns 1 and 2, a linear scale from 0 to 8 is used to describe the condition of the cutting structure according to the following:

Steel Tooth Bits
A measure of lost tooth height resulting from abrasion and/or damage
0 - No Loss of Tooth Height
8 - Total Loss of Tooth Height

Inner Bits
A measure of total cutting-structure reduction resulting from lost, worn and/or broken inserts
0 - No Lost, Worn, and/or Broken Inserts
8 - All Inserts Lost, Worn, and/or Broken

Fixed Cutter Bits
A measure of lost, worn, and/or broken cutting structure
0 - No Lost, Worn, and/or Broken Cutting Structure
8 - All of Cutting Structure Lost, Worn, and/or Broken

Cutting Structure		Location	Bearings/Seals	Gage	Other Dull Characteristics	Reason Pulled
Inner	Outer	Dull Characteristics				
1	2	3	4	5	6	7	8

Dull Characteristics (3)
(Use only cutting structure related codes)

*BF - Broken Cone
BF - Bond Failure
BT - Broken Teeth / Cutters
BU - Balled-Up Bit
*CC - Cracked Cone
*CD - Coned Dragged
CI - Cone Interference
CR - Cored
CT - Chipped Teeth / Cutters
ER - Erosion
FC - Flat-Crested Wear
HC - Heat Checking
JD - Junk Damage
*LC - Lost Cone
LN - Lost Nozzle
LT - Lost Teet / Cutters

OC - Off-Center Wear
PB - Pinched Bit
PN - Plugged Nozzle / Flow Passage
RG - Rounded Gage
RO - Ring Out
SD - Shirttail Damage
SS - Self-Sharpening Wear
TR - Tracking
WO - Washed Out Bit
WT - Worn Teeth / Cutters
NO - No Dull Characteristics

*Show cone number(s) under location

Location (4)

Roller Cone
N - Nose Row
M - Middle Row
G - Gage Row
A - All Rows
Cone #
1 2 3

Fixed Cutter
C - Cone
N - Nose
T - Taper
S - Shoulder
G - Gage
A - All Areas

Bearings/Seals (5)

Nonsealed Bearings
A liner scale estimating bearing life used
0 - No Life Used
9 - All Life Used (i.e, no bearing life remaining)

Sealed Bearings
E - Seals Effective
F - Seals Failed
N - Not Able to Grade
X - Fixed Cutter Bit

Gage (6)
Measure to nearest 1/16 of an inch
I - In Gage
1 - 1/16 in. Out of Gage
2 - 2/16 in. Out of Gage
4 - 4/16 in. Out of Gage

Other Dull Characteristics (7)
Refer to column 3 codes

Reason Pulled or Run Terminated (8)

BHA - Change BHA
DMF - Downhole Motor Failure
DTF - Downhole Tool Failure
DSF - Drillstring Failure
DST - Drillstem Test
DP - Drill Plug
CM - Condition Mud
CP - Core Point
FM - Formation Change
HP - Hole Problems

LIH - Left in Hole
HR - Hours on Bit
LOG - Run Logs
PP - Pump Pressure
PR - Penetration Rates
RIG - Rig Repair
TD - Total Depth/Casing Depth
TW - Twist Off
TQ - Torque
WC - Weather Conditions

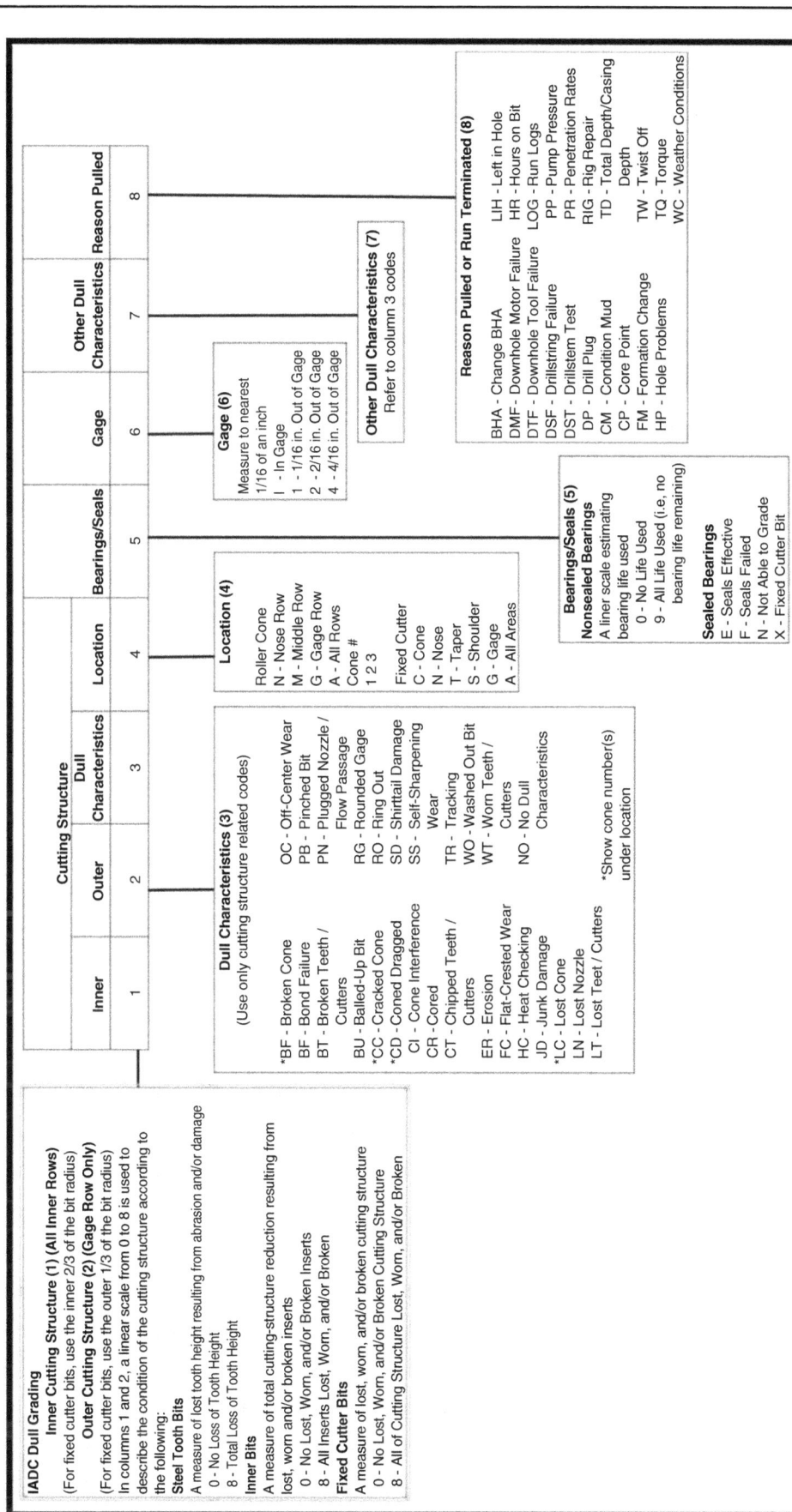

Fig. 2.10—IADC dull grading (courtesy of IADC).

Material	Specific Gravity	lbm/gal	lbm/bbl
Attapulgite	2.89	24.1	1,011
Water	1	8.33	350
Diesel	0.86	7.2	300
Bentonite clay	2.6	21.7	910
Sand	2.63	21.9	920
Average drill solids	2.6	21.7	910
API barite	4.2	35	1,470
$CaCl_2$	1.96	16.3	686
NaCL	2.16	18	756
$CaCO_3$	2.71	22.6	949
Steel	7.86	65.5	2,751

Table 2.17—Material properties (Bourgoyne et al. 1986j).

Chapter 3

Formation Evaluation

3.1 Archie's Water Saturation

$$S_w = \left(\frac{aR_w}{\phi^m R_t} \right)^{1/n}$$

S_w Water saturation
R_w Formation water resistivity, $\Omega \cdot m$
R_t Observed bulk resistivity, $\Omega \cdot m$
ϕ Porosity, unitless
a Tortuosity constant
n Saturation exponent
m Cementation exponent

Resistivity Index

$$I_R = \frac{R_t}{R_o}$$

R_t Observed bulk resistivity, $\Omega \cdot m$
R_o Resistivity of rock fully saturated with water, $\Omega \cdot m$

3.2 Formation Resistivity Factor

$$F_R = \frac{R_o}{R_w}$$

F_R Formation resistivity factor
R_o Resistivity of rock fully saturated with water, $\Omega \cdot m$
R_w Water resistivity, $\Omega \cdot m$

Formation Resistivity Factor and Porosity

$$F_R = \frac{a}{\phi^m}$$

F_R	Formation resistivity factor
ϕ	Porosity, unitless
m	Cementation constant
a	Tortuosity constant

3.3 Flushed-Zone Water Saturation

$$S_{xo} = \left(\frac{aR_{mf}}{\phi^m R_{xo}} \right)^{1/n}$$

S_{xo}	Water saturation of the flushed zone
R_{mf}	Resistivity of the mud filtrate at formation temperature, $\Omega \cdot m$
R_{xo}	Shallow resistivity, $\Omega \cdot m$
a	Tortuosity constant
n	Saturation exponent
ϕ	Porosity, unitless
m	Cementation constant

3.4 Porosity Calculations From Sonic Data

Wyllie Time Average

$$\phi_s = \frac{\Delta t_{log} - \Delta t_{matrix}}{\Delta t_f - \Delta t_{matrix}}$$

Raymer-Hunt-Gardner

$$\phi_s = \frac{5}{8} \times \left(\frac{\Delta t_{log} - \Delta t_{matrix}}{\Delta t_{log}} \right)$$

For Unconsolidated Formations

$$\phi_s = \left(\frac{\Delta t_{log} - \Delta t_{matrix}}{\Delta t_f - \Delta t_{matrix}} \right) \times \frac{1}{C_p} \qquad C_p = \frac{\Delta t_{sh} \times C}{100}$$

ϕ_s	Sonic porosity, fraction
Δt_{log}	Measured-interval travel time from sonic log, $\mu s/ft$
Δt_{matrix}	Interval travel time of the rock matrix, $\mu s/ft$
Δt_f	Interval travel time of the saturating fluid, $\mu s/ft$
Δt_{sh}	Interval travel time of the shale beds, $\mu s/ft$
C_p	Compaction factor
C	Constant, normally $= 1$

3.5 Young's Modulus

Basic Equation

$$E = \frac{9K\rho v_s^2}{3K + \rho v_s^2}$$

Equation in Well Logging Terms

$$E = \left(\frac{\rho}{\Delta t_s^{\,2}}\right)\left(\frac{3\Delta t_s^{\,2} - 4\Delta t_c^{\,2}}{\Delta t_s^{\,2} - \Delta t_c^{\,2}}\right) \times 13,400$$

E Young's modulus, psi
K Bulk modulus, psi
ρ Bulk density, lbm/ft³
v_s Shear wave velocity, ft/μs
Δt_s Shear sonic log travel time, μs/ft
Δt_c Compressional sonic log travel time, μs/ft

3.6 Bulk Modulus

Basic Equation

$$K = \rho\left(v_p^{\,2} - \tfrac{4}{3}v_s^{\,2}\right)$$

v_p Compressional or p-wave velocity, ft/μs (this unit is feet per microsecond; same units as v_s above)

Equation in Well Logging Terms

$$K = \rho\left(\frac{3\Delta t_s^{\,2} - 4\Delta t_c^{\,2}}{3\Delta t_s^{\,2} - \Delta t_c^{\,2}}\right) \times 13,400$$

K Bulk modulus, psi
ρ Bulk density, lbm/ft³
v_s Shear wave velocity, ft/μs
Δt_s Shear sonic log travel time, μs/ft
Δt_c Compressional sonic log travel time, μs/ft

3.7 Shear Modulus

Basic Equation

$$\mu = \rho v_s^{\,2}$$

Equation in Well Logging Terms

$$\mu = \frac{\rho}{\Delta t_s^{\,2}} \times 13,400$$

μ Shear modulus, psi
ρ Bulk density, lbm/ft³
v_s Shear wave velocity, ft/μs
Δt_s Shear sonic log travel time, μs/ft
Δt_c Compressional sonic log travel time, μs/ft

3.8 Poisson's Ratio

Basic Equation

$$\sigma = \frac{1}{2}\,\frac{\left(\dfrac{v_p^{\,2}}{v_s^{\,2}}\right) - 2}{\left(\dfrac{v_p^{\,2}}{v_s^{\,2}}\right) - 1}$$

σ Poisson's ratio, unitless

Equation in Well Logging Terms

$$\sigma = \frac{1}{2}\left(\frac{\Delta t_s{}^2 - 2\Delta t_c{}^2}{\Delta t_s{}^2 - \Delta t_c{}^2} \right)$$

v_s	Shear wave velocity, ft/s
v_p	Compressional wave velocity, ft/s
Δt_s	Shear sonic log travel time, µs/ft
Δt_c	Compressional sonic log travel time, µs/ft

3.9 D-Exponent

$$d_{exp} = \frac{\log\left(\dfrac{R}{60N} \right)}{\log\left(\dfrac{12W}{1000d_b} \right)}$$

Corrected D-Exponent

$$d_c = d_{exp}\frac{\rho_n}{\rho_e}$$

d_{exp}	D-exponent, in.
d_c	Corrected D-exponent, in.
R	Penetration rate, ft/hr
N	Rotary speed, rpm
W	Weight on bit, 1,000 lbf
d_b	Bit diameter, in.
ρ_n	Mud density equivalent to a normal formation pore pressure
ρ_e	Equivalent mud density at the bit while circulating

3.10 Porosity/Bulk Density

$$\phi = \frac{\rho_{ma} - \rho_b}{\rho_{ma} - \rho_{fl}}$$

ϕ	Porosity, unitless
ρ_{ma}	Matrix density, g/cm³
ρ_b	Formation bulk density, g/cm³
ρ_{fl}	Fluid density, g/cm³

Porosity From Core Analysis

$$\phi_c = \phi_e + V_{cl}\phi_{cl} = \phi_e + V_{sh}\phi_{sh}$$

ϕ_c	Core porosity
ϕ_e	Effective porosity
ϕ_{cl}	Clay porosity
ϕ_{sh}	Shale porosity
V_{cl}	Clay content
V_{sh}	Shale content

Effective Porosity

$$\phi_e = \frac{\phi_D \times \phi_{N_{sb}} - \phi_N \times \phi_{D_{sb}}}{\left(\phi_{N_{sb}} - \phi_{D_{sb}} \right) \times \left(\phi_N - \phi_D \right)}$$

ϕ_D Density porosity

ϕ_N Neutron porosity

$\phi_{D_{sb}}$ Density porosity of extraneous materials

$\phi_{N_{sb}}$ Neutron porosity of extraneous materials

3.11 Saturation

Clean Sand Model

$$S_w = \left(\frac{R_o}{R_t} \right)^{1/n}$$

S_w Water saturation, unitless

n Saturation exponent

R_t Observed bulk resistivity, $\Omega \cdot m$

R_o Resistivity of rock fully saturated with water, $\Omega \cdot m$

Laminated Sand/Shale Model

$$\frac{1}{R_t} = \frac{V_{sh}}{R_{sh}} + \frac{1 - V_{sh}}{R_{sd}}$$

$$R_{sd} = \frac{R_w}{\phi_{sd}^m S_{wsd}^n}$$

$$\phi_e = \phi_{sd} \left(1 - V_{sh} \right)$$

$$S_{we} = S_{wsd}$$

R_t True resistivity of uninvaded deep formation, $\Omega \cdot m$

R_{sh} Shale resistivity, $\Omega \cdot m$

R_{sd} Clean sand resistivity, $\Omega \cdot m$

R_w Connate-brine resistivity, $\Omega \cdot m$

V_{sh} Shale content, %bulk volume (BV)

ϕ_e Effective porosity, %BV

ϕ_{sd} Sand porosity, %BV

S_{we} Water saturation of the effective porosity, %pore volume (PV)

S_{wsd} Sand-water saturation, %PV

n Saturation exponent

m Cementation exponent

Poupon-Leveaux (Indonesia) Model

$$S_w = \left\{ \left[\left(\frac{V_{sh}^{2-V_{sh}}}{R_{sh}} \right)^{1/2} + \left(\frac{\phi_e^m}{R_w} \right)^{1/2} \right]^2 R_t \right\}^{-1/n}$$

S_w Water saturation, %PV

V_{sh} Shale content, %BV

R_{sh} Shale resistivity, $\Omega \cdot m$

R_w Connate-brine resistivity, $\Omega \cdot m$

R_t True resistivity of uninvaded, deep formation, $\Omega \cdot m$

ϕ_e Effective porosity, %BV

m Cementation exponent

n Saturation exponent

Waxman-Smits-Thomas Model

$$\frac{1}{R_t} = \phi_t^{m^*} S_{wt}^{n^*} \left(\frac{1}{R_w} + \frac{BQ_V}{S_{wt}} \right)$$

R_t True resistivity of uninvaded, deep formation, $\Omega \cdot m$
ϕ_t Total porosity, %BV
S_{wt} Water saturation of the total porosity, %PV
R_w Connate-brine resistivity, $\Omega \cdot m$
B Specific cation conductance, $\left[(1/\Omega \cdot m)/(mEq/mL) \right]$
Q_V Cation-exchange capacity of total PV, mEq/mL
m^* Waxman-Smits-Thomas cementation exponent
n^* Waxman-Smits-Thomas saturation exponent

Dual-Water Model

$$\frac{1}{R_t} = \phi_t^{m_o} S_{wt}^{n_o} \left[\frac{1}{R_{wf}} + \frac{S_{wb}}{S_{wt}} \left(\frac{1}{R_{wb}} - \frac{1}{R_{wf}} \right) \right]$$

$$\phi_e = \phi_t (1 - S_{wb})$$

$$S_{we} = \frac{S_{wt} - S_{wb}}{1 - S_{wb}}$$

R_t True resistivity of uninvaded, deep formation, $\Omega \cdot m$
R_{wb} Clay-bound water resistivity, $\Omega \cdot m$
R_{wf} Free-formation-water resistivity, $\Omega \cdot m$
ϕ_t Total porosity, %BV
ϕ_e Effective porosity, %BV
S_{wt} Water saturation of the total porosity, %PV
S_{wb} Saturation of clay-bound water in the total porosity, %PV
S_{we} Water saturation of the effective porosity, %PV
m_o Dual-water-model cementation exponent
n_o Dual-water-model saturation exponent

Saturation From Neutron Logs

$$S_w = \frac{\Sigma_{log} - \Sigma_{ma} + \phi(\Sigma_{ma} - \Sigma_h) + V_{sh}(\Sigma_{ma} - \Sigma_{sh})}{\phi(\Sigma_w - \Sigma_h)}$$

S_w Water saturation
V_{sh} Shale content, fraction
Σ_{log} Capture cross section of log response, cm^2/cm^3 or capture units
Σ_{ma} Capture cross section of rock, cm^2/cm^3 or capture units
Σ_h Capture cross section of hydrocarbon, cm^2/cm^3 or capture units
Σ_{sh} Capture cross section of shale, cm^2/cm^3 or capture units
Σ_w Capture cross section of water, cm^2/cm^3 or capture units
ϕ Porosity, unitless

Saturation From Spontaneous-Potential (SP) Logs

$$(E_{SP})_{QL} = (E_{SP})_{log} - K \log (S_w / S_{xo})^n$$

$(E_{SP})_{QL}$ Quick-look spontaneous potential, mV
$(E_{SP})_{log}$ Log spontaneous potential, mV

S_w Water saturation
S_{xo} Flushed-zone saturation
n Saturation exponent
K Coefficient, $=61.3+0.133T_f$
T_f Formation temperature, °F

Fertl and Hammack Equation

$$S_w = \left(\frac{0.81R_w}{\phi^2 R_t}\right)^{\frac{1}{2}} - \left(\frac{V_{sh}R_w}{0.4\phi R_{sh}}\right)$$

S_w Water saturation
R_w Resistivity of water, $\Omega\cdot m$
R_t Total resistivity, $\Omega\cdot m$
ϕ Porosity, unitless
V_{sh} Shale content, fraction
R_{sh} Shale resistivity, $\Omega\cdot m$

3.12 Skin Calculations

Skin Pressure Drop and Skin Factor

$$\Delta p_s = 141.2\frac{qB\mu}{kh}\left(\frac{k}{k_s}-1\right)\ln\left(\frac{r_s}{r_w}\right)$$

$$s = \left(\frac{k}{k_s}-1\right)\ln\left(\frac{r_s}{r_w}\right) = \frac{kh\Delta p_s}{141.2qB\mu}$$

Δp_s Additional pressure drop because of skin, psi
s Skin factor, dimensionless
q Flow rate at surface, STB/D
B Formation volume factor, res vol/surface vol
μ Viscosity, cp
h Net formation thickness, ft
k Permeability, md
k_s Permeability of damaged zone, md
r_w Effective wellbore radius, ft
r_s Outer radius of the damaged zone, ft

Skin With Apparent Wellbore Radius

$$s = -\ln\left(r_{wa}/r_w\right)$$

s Skin factor, dimensionless
r_{wa} Apparent or effective wellbore radius, ft
r_w Effective wellbore radius, ft

Skin in Incompletely Perforated Interval

$$s = \left(h/h_p\right)s_d + s_p$$

$$s_p = \left(\frac{1}{h_{pD}}-1\right)\ln\frac{\pi}{2r_D} + \frac{1}{h_{pD}}\ln\left[\frac{h_{pD}}{2+h_{pD}}\left(\frac{A-1}{B-1}\right)^{\frac{1}{2}}\right]$$

$$A = \frac{1}{h_{1D} + h_{pD}/4}$$

$$h_{pD} = h_p / h$$

$$h_{1D} = h_1 / h$$

$$B = \frac{1}{h_{1D} + 3h_{pD}/4}$$

s	Skin factor, dimensionless
s_d	Skin caused by formation damage, dimensionless
s_p	Skin resulting from an incompletely perforated interval, dimensionless
h	Net formation thickness, ft
h_p	Perforated interval thickness, ft
h_1	Distance from top of formation to top of perforations, ft
h_{1D}	Dimensionless distance for top of perforations, unitless
r_D	Dimensionless radius $= \frac{r_w}{h}\sqrt{\frac{k_v}{k_H}}$
r_w	Wellbore radius, ft

Skin in Deviated Wells

$$s = s_d + s_\theta$$

$$s_\theta = \left(\frac{\theta'_w}{41}\right)^{2.06} - \left(\frac{\theta'_w}{56}\right)^{1.865} \log\left(\frac{h_D}{100}\right)$$

$$\theta'_w = \tan^{-1}\left(\sqrt{\frac{k_v}{k_h}}\tan\theta_w\right)$$

$$h_D = \frac{h}{r_w}\sqrt{\frac{k_h}{k_v}}$$

s	Skin factor, dimensionless
s_d	Skin caused by formation damage, dimensionless
s_θ	Skin resulting from well inclination, dimensionless
θ_w	Well angles from the vertical
h_D	Dimensionless parameter, unitless
r_w	Effective wellbore radius, ft
h	Net formation thickness, ft
k_h	Horizontal permeability, md
k_v	Vertical permeability, md

Gravel-Pack Skin

$$s_{gp} = \frac{khL_g}{2nk_{gp}r_p^2}$$

s_{gp}	Skin factor from Darcy flow through gravel pack, dimensionless
h	Net formation thickness, ft
k	Reservoir permeability, md
k_{gp}	Reservoir permeability, md
L_g	Length of the flow path through the gravel pack, ft
n	Number of perforations open
r_p	Radius of the perforation tunnel, ft

Completion Skin

$$s = s_d + s_p + s_{dp}$$

$$s_{dp} = \left(\frac{h}{L_p n}\right)\left(\ln\frac{r_{dp}}{r_p}\right)\left(\frac{k}{k_{dp}} - \frac{k}{k_d}\right)$$

s	Completion skin, unitless
s_d	Skin caused by formation damage, dimensionless
s_p	Skin resulting from an incompletely perforated interval, dimensionless
s_{dp}	Perforation damage skin, dimensionless
h	Net formation thickness, ft
k	Reservoir permeability, md
k_d	Permeability of the damaged zone around the wellbore, md
k_{dp}	Permeability of the damaged zone around perforation tunnels, md
L_p	Length of perforation tunnel, ft
n	Number of perforations open
r_p	Radius of the perforation tunnel, ft
r_{dp}	Radius of the damaged zone around the perforation tunnel, ft

Non-Darcy Flow Skin

$$s' = s + D|q_g|$$

s'	Apparent skin factor, dimensionless
s	True skin because of damage or stimulation, dimensionless
D	Non-Darcy flow coefficient, D/Mscf
q_g	Gas flow rate, Mscf/D

Skin Factor for Multiphase-Flow Test Analysis Using Semilog Plots

$$s = 1.151\left[\frac{\Delta p_{1\,hr}}{-m} - \log\left(\frac{\lambda_t}{\phi c_t r_w^2}\right) + 3.23\right]$$

$$\lambda_t = \frac{k_o}{\mu_o} + \frac{k_w}{\mu_w} + \frac{k_g}{\mu_g}$$

$\Delta p_{1\,hr}$	Pressure change from start of test to one hour elapsed time, psi
m	Slope of middle-time line, psi/cycle
λ_t	Total mobility, md/cp
k_o	Permeability of oil, md
k_w	Permeability of gas, md
k_g	Permeability of water, md
μ_o	Oil viscosity, cp
μ_g	Gas viscosity, cp
μ_w	Water viscosity, cp
ϕ	Porosity, unitless
c_t	Total compressibility, psi^{-1}
r_w	Effective wellbore radius, ft

Fracture Damage

Chocked-Fracture Skin

$$s = \frac{\pi k L_s}{k_{fs} w_f}$$

k	Reservoir permeability, md
k_{fs}	Reduced permeability near wellbore, md
L_s	Length of damaged zone in fracture, ft
w_f	Fracture width, ft

Fracture-Face Skin

$$s = \frac{\pi w_s}{2L_f}\left(\frac{k}{k_s} - 1\right)$$

k	Reservoir permeability, md
k_s	Permeability of altered zone, md
L_f	Fracture half-length, ft
w_s	Width of damaged zone around fracture face, ft

3.13 Horizontal Flow Influx Equations

Early Radial Flow

$$p_i - p_{wf} = \frac{162.6qB\mu}{\sqrt{k_x k_z}L_w}\left[\log_{10}\left(\frac{\sqrt{k_x k_z}\,t}{\phi\mu c_t r_w^2}\right) - 3.227 + 0.868 s_d\right]$$

End of Early Radial Flow

$$t_{Eerf} = \frac{1{,}800 d_z^2 \phi\mu c_t}{k_z}$$

$$t_{Eerf} = \frac{125 L_w^2 \phi\mu c_t}{k_y}$$

t_{Eerf}	End of early radial flow, hours

Skin for Early Radial Flow

$$s_d = 1.151\left[\frac{\Delta p_{1\,hr}}{|m_{erf}|} - \log\left(\frac{\sqrt{k_x k_z}}{\phi c_t r_w^2}\right) + 3.23\right]$$

p_i	Initial reservoir pressure, psi
p_{wf}	Flowing bottomhole pressure, psi
q	Flow rate at surface, STB/D
B	Formation volume factor, res vol/surface vol
μ	Viscosity, cp
k_x	Permeability in x-direction, md
k_y	Permeability in y-direction, md
k_z	Permeability in z-direction, md
L_w	Completed length of horizontal well, ft
d_z	Shortest distance between horizontal well and z boundary, ft
t	Elapsed time, hours
c_t	Total compressibility, psi^{-1}
r_w	Effective wellbore radius, ft
s_d	Skin caused by formation damage, dimensionless
$\Delta p_{1\,hr}$	Pressure change from start of test to one hour elapsed time, psi
m_{erf}	Slope of semilog plot of early radial flow, psi/log cycle
ϕ	Porosity, unitless

Hemi-Radial Flow

$$p_i - p_{wf} = \frac{325.2qB\mu}{\sqrt{k_x k_z}L_w}\left[\log_{10}\left(\frac{\sqrt{k_x k_z}\,t}{\phi c_t r_w^2}\right) - 3.227 + 0.868 s_d\right]$$

Start of Hemi-Radial Flow

$$t_{Shrf} \approx \frac{1,800 d_z^2 \phi \mu c_t}{k_z}$$

t_{Shrf} Start of hemi-radial flow, hours

End of Hemi-Radial Flow

$$t_{Eerf} = \frac{1,800 D_z^2 \phi \mu c_t}{k_z}$$

$$t_{Eerf} = \frac{125 L_w^2 \phi \mu c_t}{k_y}$$

Skin for Hemi-Radial Flow

$$s_d = 2.303 \left[\frac{\Delta p_{1\,hr}}{|m_{hrf}|} - \log\left(\frac{\sqrt{k_x k_z}}{\phi \mu c_t r_w^2} \right) + 3.23 \right]$$

p_i Initial reservoir pressure, psi
p_{wf} Flowing bottomhole pressure, psi
q Flow rate at surface, STB/D
B Formation volume factor, res vol/surface vol
μ Viscosity, cp
k_x Permeability in x-direction, md
k_y Permeability in y-direction, md
k_z Permeability in z-direction, md
L_w Completed length of horizontal well, ft
d_z Shortest distance between horizontal well and z boundary, ft
D_z Longest distance between horizontal well and z boundry, ft
t Elapsed time, hours
c_t Total compressibility, psi^{-1}
r_w Effective wellbore radius, ft
s_d Skin caused by formation damage, dimensionless
$\Delta p_{1\,hr}$ Pressure change from start of test to one hour elapsed time, psi
m_{hrf} Slope of semilog plot of hemi-radial flow, psi/log cycle
ϕ Porosity, unitless

Early Linear Flow

$$p_i - p_{wf} = \frac{8.128 qB}{hL_w} \sqrt{\frac{\mu t}{k_x \phi c_t}} + \frac{141.2 qB\mu}{L_w \sqrt{k_x k_z}} (s_c + s_d)$$

Start of Early Linear Flow

$$t_{Self} = \frac{1,800 D_z^2 \phi \mu c_t}{k_z}$$

t_{Self} End of hemi-radial flow, hours

End of Early Linear Flow

$$t_{Eelf} = \frac{160 L_w^2 \phi \mu c_t}{k_y}$$

t_{Eelf} End of early linear flow, hours

Horizontal Permeability Perpendicular to the Well

$$\sqrt{k_x} = \frac{8.128qB}{|m_{\text{elf}}|\,L_w h}\sqrt{\frac{\mu}{\phi c_t}}$$

Altered-Permeability Skin for Early Linear Flow

$$s_d = \frac{L_w\sqrt{k_x k_z}}{141.2qB\mu}\Delta p_{t=0} - s_c$$

$$s_c = \ln\left(\frac{h}{r_w}\right) + 0.25\ln\left(\frac{k_x}{k_z}\right) - \ln\left[\sin\left(\frac{\pi d_z}{h}\right)\right] - 1.838$$

$\Delta p_{t=0}$ $p_i - p_{wf}$, Difference between initial reservoir pressure and flowing bottomhole pressure, psi
p_i Initial reservoir pressure, psi
p_{wf} Flowing bottomhole pressure, psi
q Flow rate at surface, STB/D
B Formation volume factor, res vol/surface vol
μ Viscosity, cp
k_x Permeability in x-direction, md
k_y Permeability in y-direction, md
k_z Permeability in z-direction, md
L_w Completed length of horizontal well, ft
d_z Shortest distance between horizontal well and z boundary, ft
D_z Longest distance between horizontal well and z boundry, ft
t Elapsed time, hours
c_t Total compressibility, psi^{-1}
r_w Effective wellbore radius, ft
h Net formation thickness, ft
s_d Skin caused by formation damage, dimensionless
s_c Convergence skin, dimensionless
m_{elf} Slope of square-root-of-time plot for early radial flow, psi/log cycle
ϕ Porosity, unitless

Late Pseudoradial Flow

$$p_i - p_{wf} = \frac{162.6qB\mu}{\sqrt{k_x k_z}\,h}\left[\log_{10}\left(\frac{k_y t}{\phi\mu c_t L_w^2}\right) - 2.303\right] + \frac{141.2qB\mu}{L_w\sqrt{k_x k_z}}\left(s_c + s_d\right)$$

Start of Pseudoradial Flow

$$t_{Sprf} = \frac{1480 L_w^2 \phi\mu c_t}{k_y}$$

t_{Sprf} Start of pseudoradial flow, hours

End of Pseudoradial Flow

$$t_{Eprf} = \frac{2{,}000\phi\mu c_t \left(L_w/4 + d_y\right)^2}{k_y}$$

$$t_{Eprf} = \frac{1{,}650\phi\mu c_t d_x^2}{k_x}$$

t_{Eprf} End of pseudoradial flow, hours

Permeability in Horizontal Plane

$$\sqrt{k_x k_y} = \frac{162.6 q B \mu}{|m_{\text{prf}}| h}$$

h Net formation thickness, ft

Altered Permeability Skin for Pseudoradial Flow

$$s_d = \left(1.151 \sqrt{\frac{k_z}{k_y}} \frac{L_w}{h}\right) \left[\frac{\Delta p_{1\,\text{hr}}}{|m_{\text{prf}}|} - \log\left(\frac{k_y}{\phi \mu c_t L_w^2}\right) + 1.76\right] - s_c$$

p_i Initial reservoir pressure, psi
p_{wf} Flowing bottomhole pressure, psi
q Flow rate at surface, STB/D
A Drainage area, ft^2
B Formation volume factor, res vol/surface vol
μ Viscosity, cp
k_x Permeability in x-direction, md
k_y Permeability in y-direction, md
k_z Permeability in z-direction, md
L_w Completed length of horizontal well, ft
d_x Shortest distance between horizontal well and x boundary, ft
d_y Shortest distance between horizontal well and y boundary, ft
t Elapsed time, hours
c_t Total compressibility, psi^{-1}
r_w Effective wellbore radius, ft
h Net formation thickness, ft
s_d Skin caused by formation damage, dimensionless
s_c Convergence skin, dimensionless
$\Delta p_{1\,\text{hr}}$ Pressure change from start of test to one hour elapsed time, psi
m_{prf} Slope of semilog plot for pseudoradial flow, psi/log cycle
ϕ Porosity, unitless

Late Linear Flow

$$p_i - p_{wf} = \frac{8.128 q B}{h b_H} \sqrt{\frac{\mu t}{k_x \phi c_t}} + \frac{141.2 q B \mu}{b_H \sqrt{k_x k_z}} \left(s_p + s_c + \frac{b_H}{L_w} s_d\right)$$

Start of Late Linear Flow

$$t_{\text{Sllf}} = \frac{4{,}800 \phi \mu c_t \left(d_y + L_w / 4\right)^2}{k_y}$$

$$t_{\text{Sllf}} = \frac{1{,}800 \phi \mu c_t D_z^2}{k_z}$$

t_{Sllf} Start of late linear flow, hours

End of Late Linear Flow

$$t_{\text{Ellf}} = \frac{1{,}650 \phi \mu c_t d_x^2}{k_x}$$

t_{Ellf} End of late linear flow, hours

Length of Drainage Area

$$b_H = \frac{8.128qB}{|m_{llf}|h}\sqrt{\frac{\mu}{\phi c_t k_x}}$$

Altered Permeability Skin for Late Linear Flow

$$s_d = \frac{L_w}{b_H}\left[\frac{b_H\sqrt{k_x k_z}\left(\Delta p_{t=0}\right)}{141.2qB\mu}\right] - s_c - s_p$$

$\Delta p_{t=0}$ $p_i - p_{wf}$, Difference between initial reservoir pressure and flowing bottomhole pressure, psi

p_i Initial reservoir pressure, psi

p_{wf} Flowing bottomhole pressure, psi

q Flow rate at surface, STB/D

A Drainage area, ft^2

B Formation volume factor, res vol/surface vol

μ Viscosity, cp

k_x Permeability in x-direction, md

k_y Permeability in y-direction, md

k_z Permeability in z-direction, md

L_w Completed length of horizontal well, ft

d_x Shortest distance between horizontal well and x boundary, ft

d_y Shortest distance between horizontal well and y boundary, ft

D_z Longest distance between horizontal well and z boundry, ft

t Elapsed time, hours

c_t Total compressibility, psi^{-1}

r_w Effective wellbore radius, ft

h Net formation thickness, ft

b_H Length in direction parallel to wellbore, ft

s_d Skin caused by formation damage, dimensionless

s_c Convergence skin, dimensionless

s_p Skin resulting from an incompletely perforated interval, dimensionless

m_{llf} Slope of square-root-of-time plot for late linear flow, psi/log cycle

ϕ Porosity, unitless

Uniform Flux Equation (Babu-Odeh Method)

$$q = \frac{0.00708 b_H \sqrt{k_x k_z}\left(\bar{p} - p_{wf}\right)}{B\mu\left[\ln\left(\dfrac{A^{1/2}}{r_w}\right) + \ln C_H - 0.75 + s_p + \left(\dfrac{b_H}{L_w}\right)s_d\right]}$$

$$\ln C_H = 6.28\frac{a_H}{h}\sqrt{\frac{k_z}{k_x}}\left[\frac{1}{3} - \frac{d_x}{a_H} + \left(\frac{d_x}{a_H}\right)^2\right] - \ln\left(\sin\frac{\pi d_z}{h}\right) - 0.5\ln\left[\left(a_H / h\right)\sqrt{k_z / k_x}\right] - 1.088$$

p_i Initial reservoir pressure, psi

p_{wf} Flowing bottomhole pressure, psi

$\bar{p}$ Volumetric average or static drainage-area pressure, psi

q Flow rate at surface, STB/D

A Drainage area, ft^2

B Formation volume factor, res vol/surface vol

C_H	Wellbore Storage term for a horizontal well
μ	Viscosity, cp
k_x	Permeability in x-direction, md
k_z	Permeability in z-direction, md
L_w	Completed length of horizontal well, ft
d_x	Shortest distance between horizontal well and x boundary, ft
d_z	Shortest distance between horizontal well and z boundary, ft
t	Elapsed time, hours
r_w	Effective wellbore radius, ft
h	Net formation thickness, ft
a_H	Total width of reservoir perpendicular to the wellbore, ft
b_H	Length in direction parallel to wellbore, ft
s_d	Skin caused by formation damage, dimensionless
s_p	Skin resulting from an incompletely perforated interval, dimensionless
ϕ	Porosity, unitless

Productivity Index

$$J = \frac{\bar{k}b_H}{887.22 B\mu \left(p_D + \frac{b_H}{2\pi L_w}\sum s \right)}$$

$$p_D = \frac{b_H C_H}{4\pi h} + \frac{b_H}{2\pi L_w} s_c$$

$$s_c = \ln\left(\frac{h}{2\pi r_w}\right) - \frac{h}{6L_w} + s_e$$

$$s_e = \frac{h}{L_w}\left[\frac{2d_z}{h} - \frac{1}{2}\left(\frac{2d_z}{h}\right)^2 - \frac{1}{2}\right] - \ln\left[\sin\left(\frac{\pi d_z}{h}\right)\right]$$

J	Productivity index, STB/(D psi)
p_D	Dimensionless pressure as defined for constant-rate production $= 0.00708\, kh(p_i - p)/qB\mu$
k	Matrix permeability, md
p_i	Original reservoir pressure, psi
p	Pressure, psi
q	Flow rate at surface, STB/D
μ	Viscosity, cp
$\bar{k}$	Average permeability, md
L_w	Completed length of horizontal well, ft
d_z	Shortest distance between horizontal well and z boundary, ft
r_w	Effective wellbore radius, ft
h	Net formation thickness, ft
b_H	Length in direction parallel to wellbore, ft
s_c	Convergence skin, dimensionless
s_e	Skin caused by eccentric effects, dimensionless
$\sum s$	Sum of damage skin, turbulence, and other pseudoskin factors
ϕ	Porosity, unitless
B	Formation volume factor, res vol/surface vol
C_H	Wellbore storage term for a horizontal well

Logging Charts [from Halliburton's *Log Interpretation Chart* book (Halliburton Energy Services 1994)]

SW-1 (Archie Nomograph)

Applications:	Determination of formation water saturation.
Nomenclature:	R_o Formation resistivity when formation water saturation is 100%
	R_t True formation resistivity
	R_w Formation water resistivity at formation temperature
	F Formation resistivity factor
	ϕ Formation porosity
	S_w Formation water saturation
	n Saturation exponent
	a Formation factor coefficient
	m Cementation exponent
Given:	$R_t = 9\ \Omega\cdot m$
	$R_w = 0.06\ \Omega\cdot m$
	$\phi = 20\%$
	$n = 2$
Find:	S_w
Procedure:	Enter the nomograph on the R_w leg at 0.06 $\Omega\cdot$m and project through the ϕ leg at 20%. The projection intersects the R_o leg at approximately 1.2 $\Omega\cdot$m. (The value of F corresponding to the given values of R_w and ϕ can be read off the F leg. In this example, the value of F is 20.)
	From 1.2 $\Omega\cdot$m on the R_o leg, project through 9 $\Omega\cdot$m on the R_t leg. The projection intersects the S_w^n leg at approximately 0.13.
	From 0.13 on the S_w^n leg, project through 2.0 on the n leg. The projection intersects the S_w leg at approximately 36.0%.
Answer:	$S_w = 36.0\%$
Notes:	This chart assumes $a = 0.62$ and $m = 2.15$:

$$F = \frac{a}{\phi^m}$$

You can calculate S_w in decimal form from the Archie equation:

$$S_w = \sqrt[n]{\frac{a \cdot R_w}{\phi^m \cdot R_t}}$$

In the above equations, you must use the decimal form of ϕ.

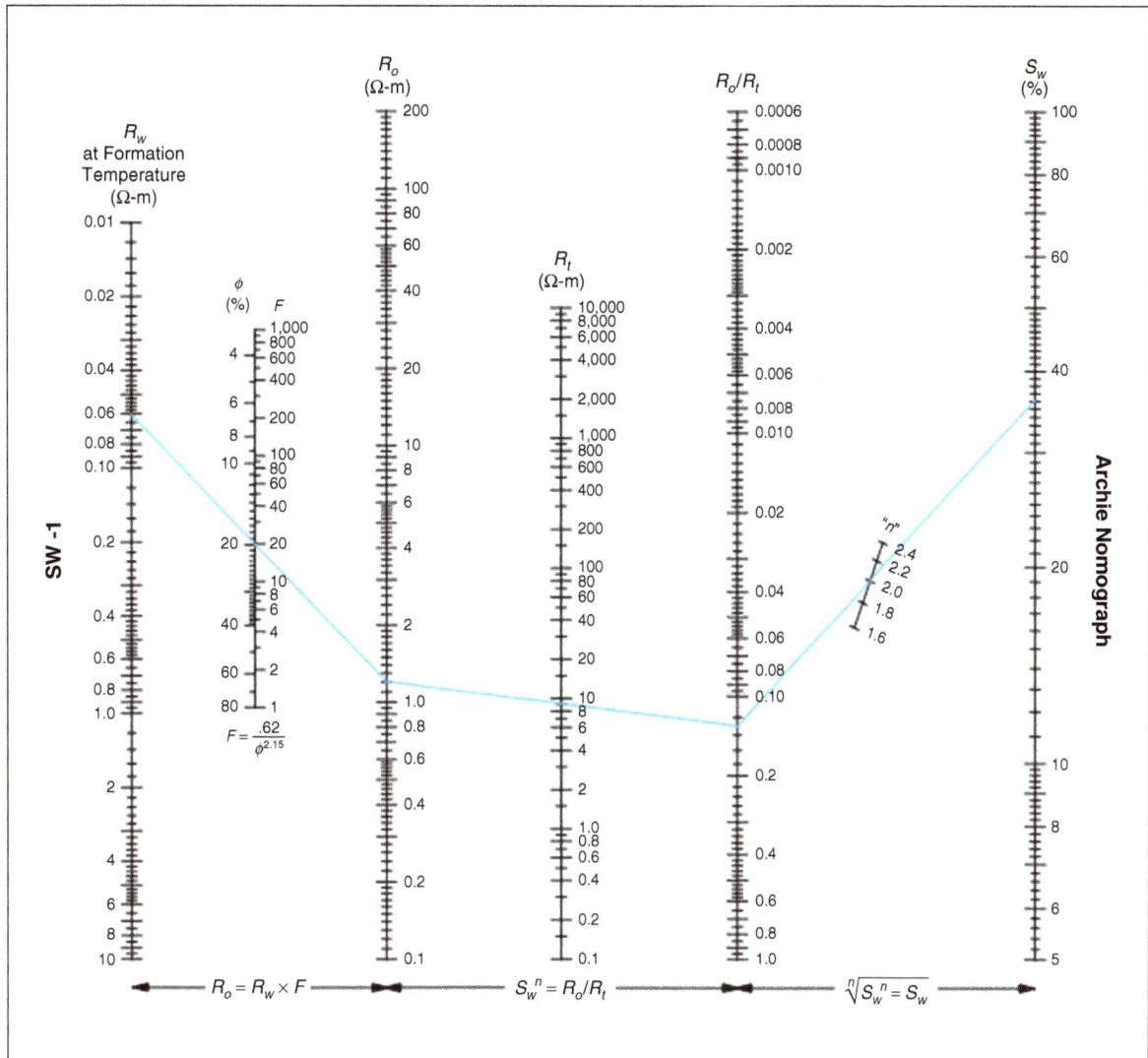

Fig. 3.1—SW-1 Archie Nomograph (Archie 1942).

Borehole Diagram With Interpretation Symbols
Open Hole

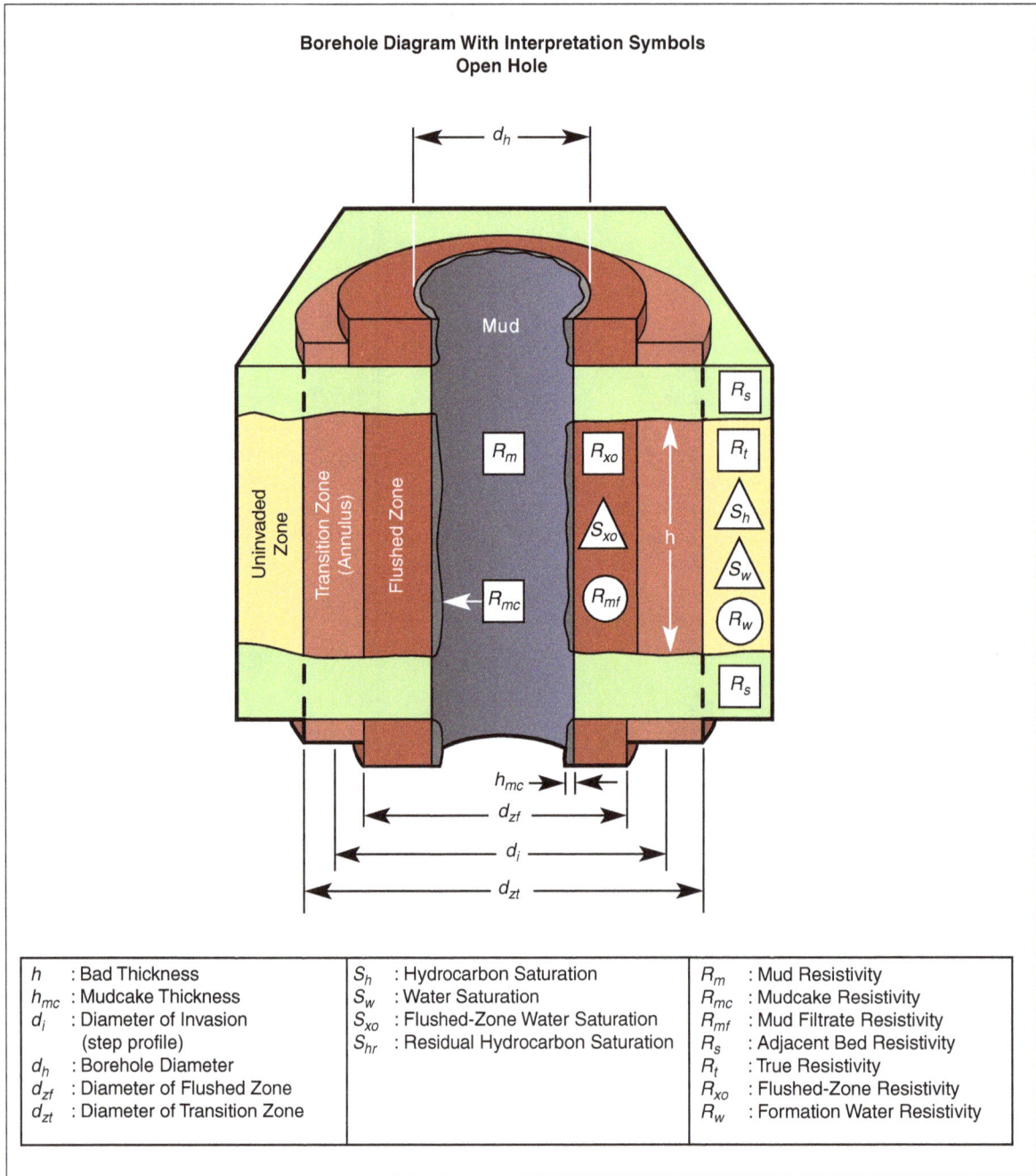

h : Bad Thickness	S_h : Hydrocarbon Saturation	R_m : Mud Resistivity
h_{mc} : Mudcake Thickness	S_w : Water Saturation	R_{mc} : Mudcake Resistivity
d_i : Diameter of Invasion (step profile)	S_{xo} : Flushed-Zone Water Saturation	R_{mf} : Mud Filtrate Resistivity
d_h : Borehole Diameter	S_{hr} : Residual Hydrocarbon Saturation	R_s : Adjacent Bed Resistivity
d_{zf} : Diameter of Flushed Zone		R_t : True Resistivity
d_{zt} : Diameter of Transition Zone		R_{xo} : Flushed-Zone Resistivity
		R_w : Formation Water Resistivity

Fig. 3.2—Gen-1a Borehole diagram with interpretation symbols, open hole (Halliburton Energy Services 1994).

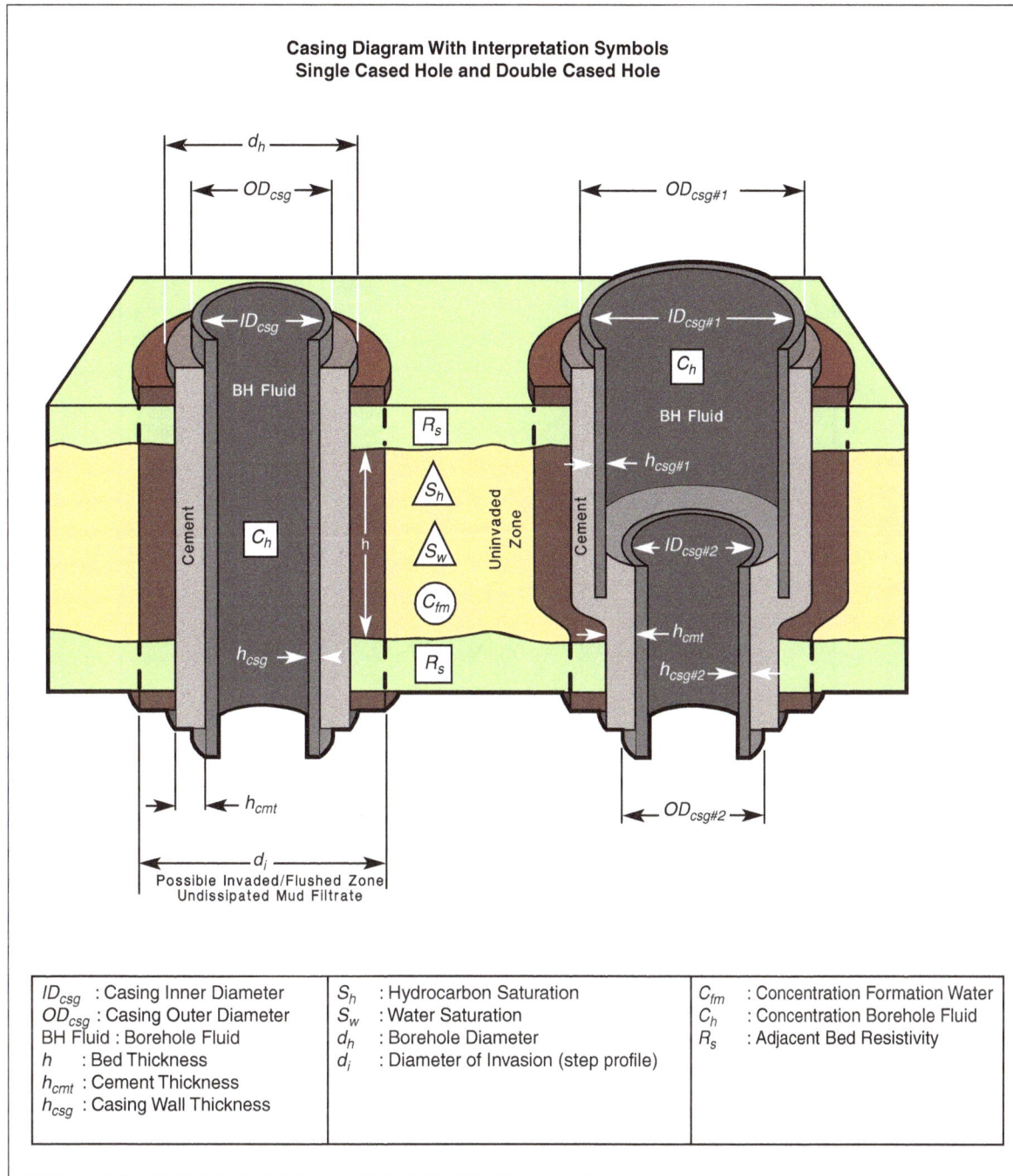

Casing Diagram With Interpretation Symbols
Single Cased Hole and Double Cased Hole

ID_{csg} : Casing Inner Diameter	S_h : Hydrocarbon Saturation	C_{fm} : Concentration Formation Water
OD_{csg} : Casing Outer Diameter	S_w : Water Saturation	C_h : Concentration Borehole Fluid
BH Fluid : Borehole Fluid	d_h : Borehole Diameter	R_s : Adjacent Bed Resistivity
h : Bed Thickness	d_i : Diameter of Invasion (step profile)	
h_{cmt} : Cement Thickness		
h_{csg} : Casing Wall Thickness		

Fig. 3.3—GEN-1b Casing diagram with interpretation symbols, single-cased hole and double-cased hole (Halliburton Energy Services 1994).

Depth – Temperature – Geothermal Gradient
International

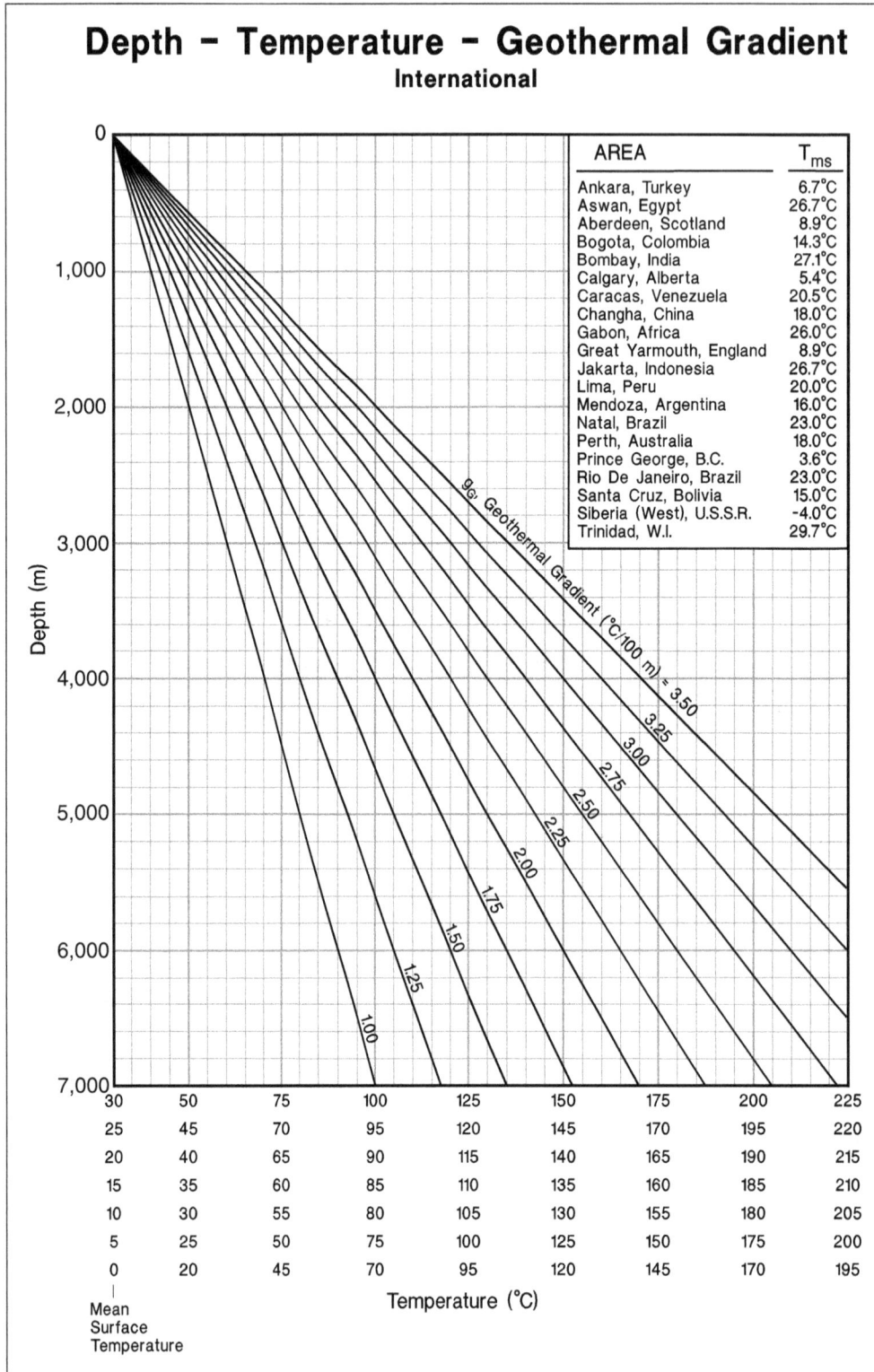

AREA	T_{ms}
Ankara, Turkey	6.7°C
Aswan, Egypt	26.7°C
Aberdeen, Scotland	8.9°C
Bogota, Colombia	14.3°C
Bombay, India	27.1°C
Calgary, Alberta	5.4°C
Caracas, Venezuela	20.5°C
Changha, China	18.0°C
Gabon, Africa	26.0°C
Great Yarmouth, England	8.9°C
Jakarta, Indonesia	26.7°C
Lima, Peru	20.0°C
Mendoza, Argentina	16.0°C
Natal, Brazil	23.0°C
Perth, Australia	18.0°C
Prince George, B.C.	3.6°C
Rio De Janeiro, Brazil	23.0°C
Santa Cruz, Bolivia	15.0°C
Siberia (West), U.S.S.R.	-4.0°C
Trinidad, W.I.	29.7°C

Fig. 3.4—GEN-2a International geothermal gradient (Halliburton Energy Services 1994).

Depth – Temperature – Geothermal Gradient
North America

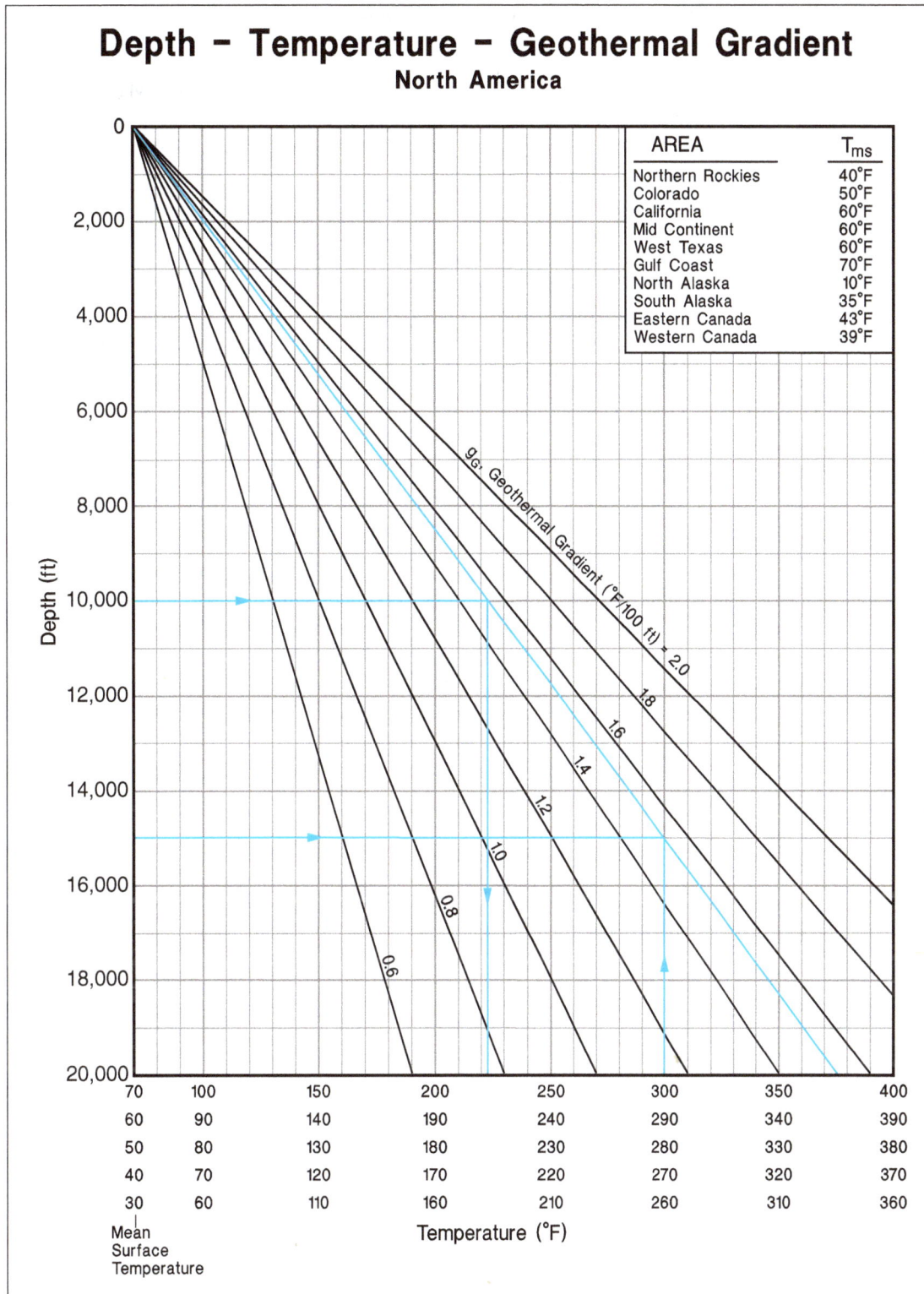

AREA	T_{ms}
Northern Rockies	40°F
Colorado	50°F
California	60°F
Mid Continent	60°F
West Texas	60°F
Gulf Coast	70°F
North Alaska	10°F
South Alaska	35°F
Eastern Canada	43°F
Western Canada	39°F

Fig. 3.5—GEN-2b North America geothermal gradient (Halliburton Energy Services 1994).

GEN-5 (Resistivity-Salinity-Temperature Conversions of NaCl Solutions; see Fig. 3.6)

Applications:

1. Determination of the resistivity of a solution at a given temperature when the solution's NaCl concentration is known and vice versa.
2. Determination of the resistivity of an NaCl solution at a given temperature when its resistivity at another temperature is known.

Nomenclature: R_w Water resistivity

Example 1

Given: Water with an NaCl concentration of 10,000 ppm

Find: R_w @ 300°F

Procedure: Enter the chart at 300°F on the lower Temperature axis. Project vertically to the 10,000-ppm NaCl curve, then horizontally to the right-hand Resistivity axis, and there estimating R_w to be 0.14 Ω·m.

Answer: $R_w = 0.14$ Ω·m @ 300°F

Example 2

Given: $R_w = 0.7$ Ω·m @ 200°F

Find: R_w @ 100°F

Procedure: Enter the chart at 0.7 Ω·m on the right-hand Resistivity axis. Project horizontally to the 200°F line. The point of intersection lies on the 3,000-ppm Salinity curve. Follow the 3,000-ppm Salinity curve until it intersects the 100°F line, then project horizontally to the left-hand Resistivity axis, and there estimating R_w to be 1.4 Ω·m.

Answer: $R_w = 1.4$ Ω·m @ 100°F

Notes: Although less accurate, you can use the following equation to approximate such temperature-induced changes in the resistivity of an NaCl solution.

$$R_2 = R_1 \left(\frac{T_1 + k}{T_2 + k} \right)$$

where $k = 6.77$ when T_1 and T_2 are expressed in °F

        $k = 21.5$ when T_1 and T_2 are expressed in °C

Resistivity-Salinity-Temperature Conversions of NaCl Solutions

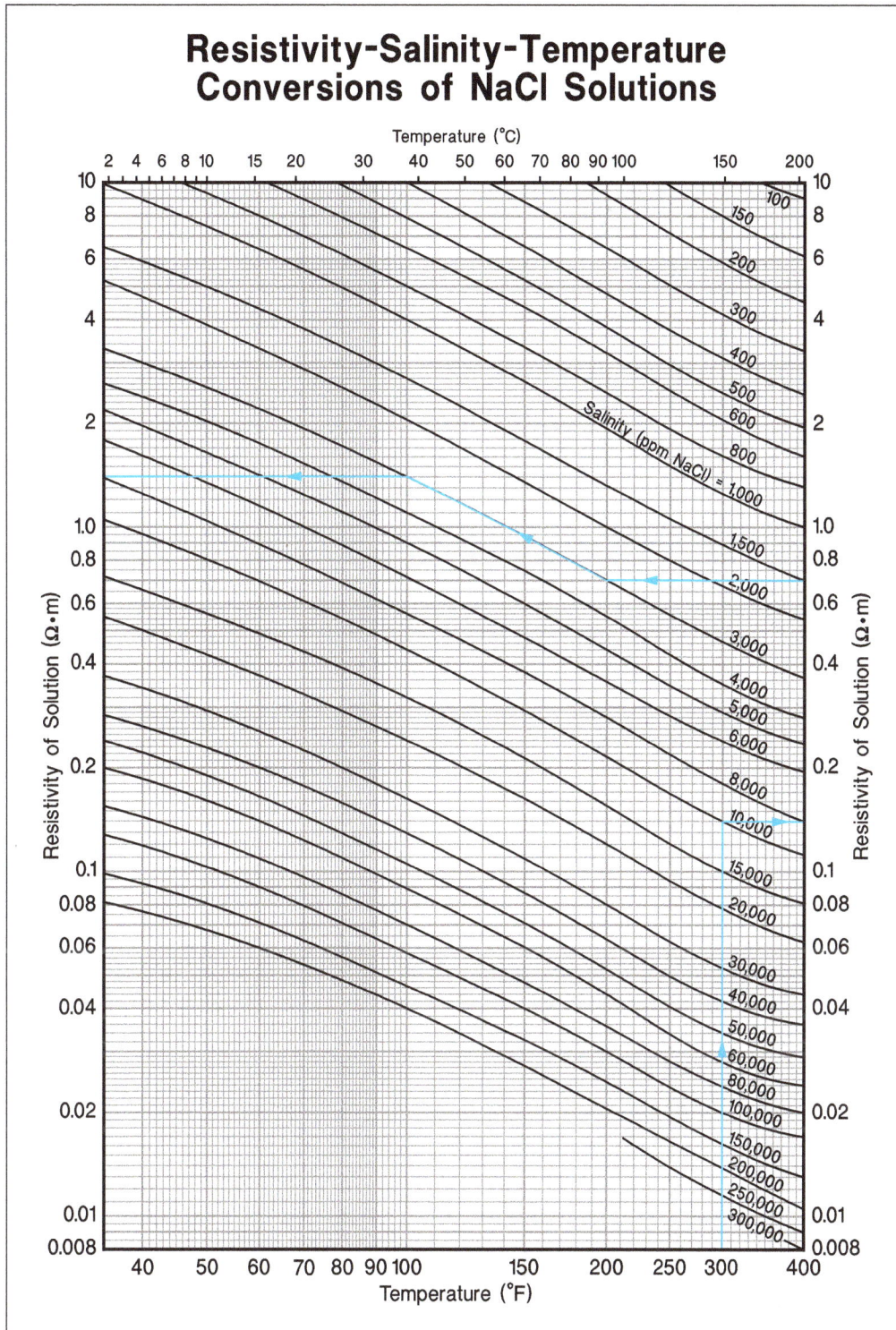

Fig. 3.6—GEN-5 Resistivity-salinity-temperature conversions of NaCl solutions (Arps 1953; Halliburton Energy Services 1994).

GR-1 (Gamma Ray Borehole Corrections)

Applications:	Correction of gamma ray measurements for borehole effects
Nomenclature:	d_t Tool diameter
	GR Gamma Ray
	GR_{cor} Gamma Ray corrected for borehole effects
	d_h Borehole diameter
	ρ_m Mud density
Given:	$d_t = 4$ in.
	Tool centered in borehole
	$GR = 90$ API units
	$d_h = 10$ in.
	$\rho_m = 12$ lb/gal
Find:	GR_{cor}

Procedure: Use the chart labeled "Tool Diameter=4 in." Enter the chart on the Hole Diameter axis at 10-in. Project vertically to the "centered" 12 lb/gal Mud Weight curve, then horizontally to the $\dfrac{GR_{cor}}{GR}$ axis, there estimating $\dfrac{GR_{cor}}{GR}$ to be 1.5. Calculate

$$GR_{cor} = \left(\frac{GR_{cor}}{GR} \right) \cdot GR = 1.5 \cdot 90 \text{ API units} = 135 \text{ API units}$$

Answer: $GR_{cor} = 135$ API units

Notes: GR_{cor} can be calculated with the following equations:

$$GR_{cor} = \left(\frac{GR_{cor}}{GR} \right) \cdot GR = 1.5 \cdot 90 \text{ API units} = 135 \text{ API units}$$

Centered or Eccentered Tool

$$GR_{cor} = GR \cdot \left[0.8 + \left(-0.10855 + 0.019 \ \rho_m \right) \cdot \left(d_h - d_t \right) \right]$$

In these equations, ρ_m is expressed in lb/gal and d_h and d_t are expressed in inches.

Gamma Ray Borehole Corrections

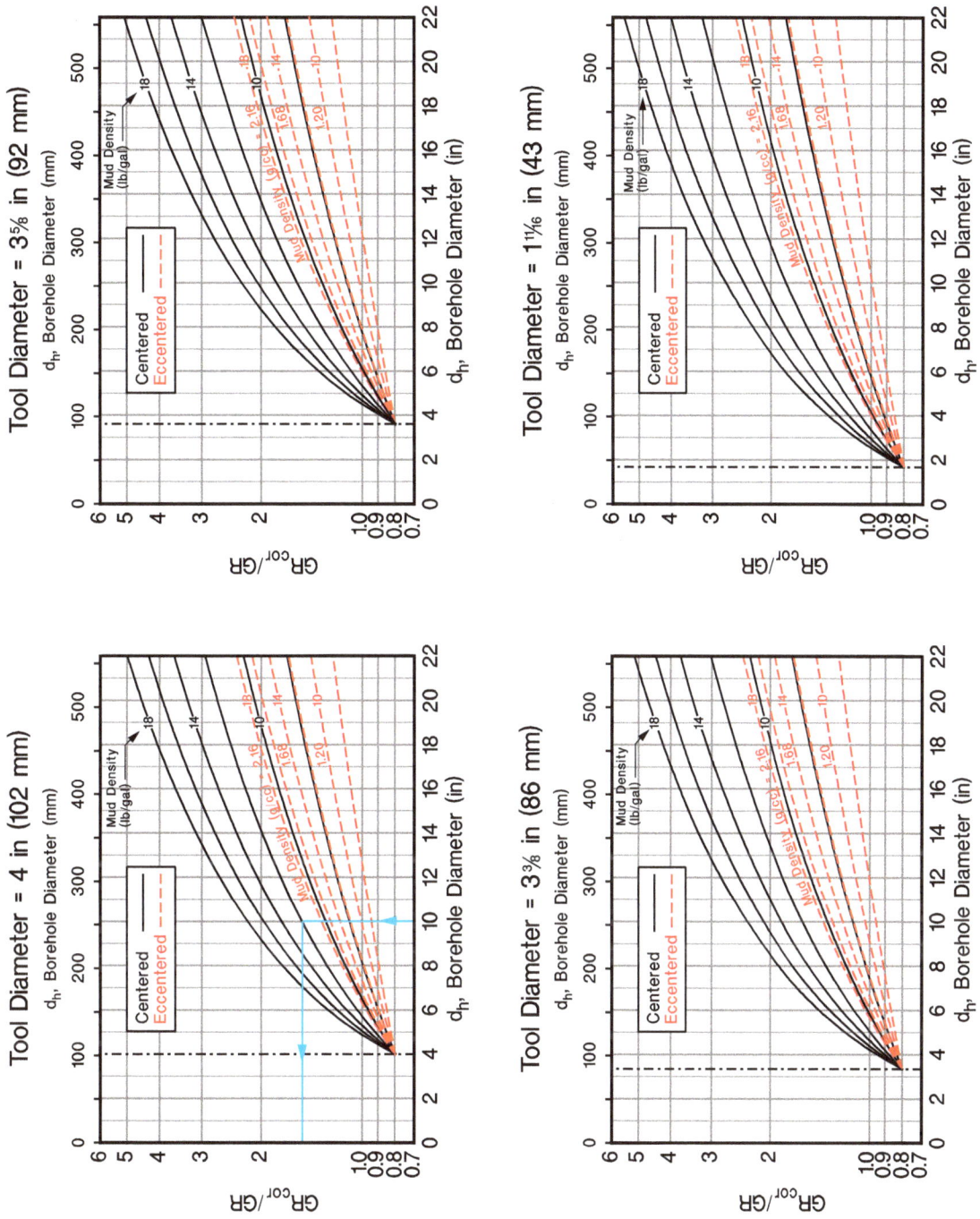

Fig. 3.7—GR-1 Gamma ray borehole corrections (Halliburton Energy Services 1994).

POR-10 [Porosity Determination: Bulk (Log) Density vs. Porosity]

Applications: Determination of formation porosity from formation bulk density

Nomenclature:
ρ_b Formation bulk density

ρ_{ma} Formation matrix density

ρ_f Formation fluid density in zone of investigation

ϕ Formation porosity

Given: Borehole fluid is fresh mud.

$\rho_b = 2.45$ g/cc (from density log corrected for borehole effects)

$\rho_{ma} = 2.71$ g/cc

$\rho_f = 1.0$ g/cc

(Estimated. This is a reasonable value for pore fluids near the wellbore when fresh muds are used.)

Find: ϕ

Procedure: Enter the chart on the Bulk Density axis at 2.45 g/cc. Project vertically to the $\rho_{ma} = 2.71$ g/cc curve, then horizontally to the $\rho_f = 1.0$ g/cc Porosity axis, there estimating ϕ to be 15.2%.

Answer: $\phi = 15.2\%$

Notes: You can calculate ϕ in decimal form from the following equation:

$$\phi = \frac{\rho_{ma} - \rho_b}{\rho_{ma} - \rho_f}$$

with $\rho_{ma}, \rho_b,$ and ρ_f in g/cc.

Porosity Determination
Bulk (Log) Density versus Porosity

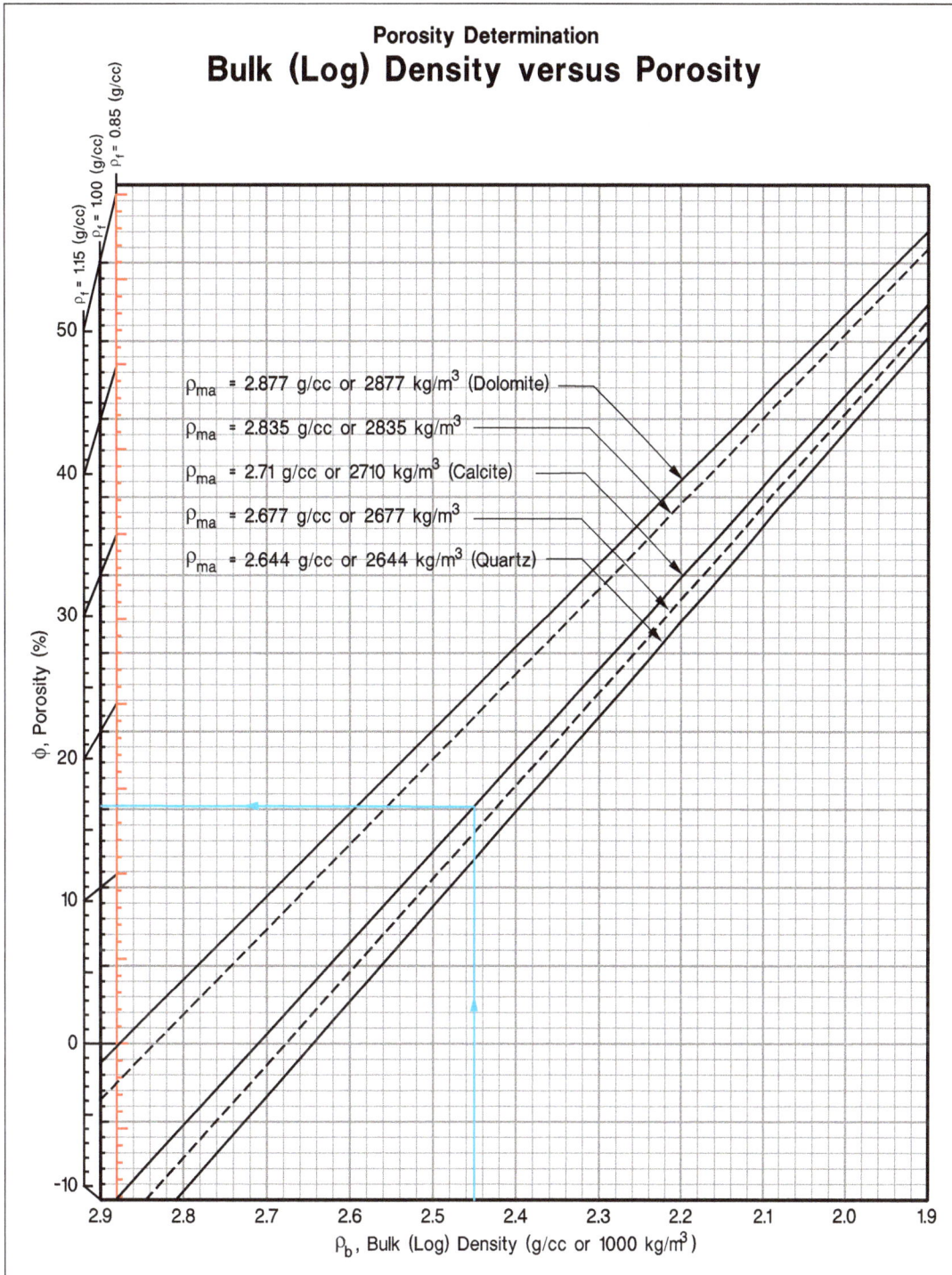

$\rho_f = 0.85$ (g/cc)
$\rho_f = 1.00$ (g/cc)
$\rho_f = 1.15$ (g/cc)

$\rho_{ma} = 2.877$ g/cc or 2877 kg/m^3 (Dolomite)

$\rho_{ma} = 2.835$ g/cc or 2835 kg/m^3

$\rho_{ma} = 2.71$ g/cc or 2710 kg/m^3 (Calcite)

$\rho_{ma} = 2.677$ g/cc or 2677 kg/m^3

$\rho_{ma} = 2.644$ g/cc or 2644 kg/m^3 (Quartz)

ϕ, Porosity (%)

ρ_b, Bulk (Log) Density (g/cc or 1000 kg/m^3)

Fig. 3.8—POR-10 Bulk (log) density vs. porosity (Halliburton Energy Services 1994).

POR-11 (Porosity Determination: Sonic vs. Porosity)

Applications:	Determination of porosity from sonic interval transit time
Nomenclature:	Δt_c Sonic compressional interval transit time in formation
	Δt_{ma} Sonic compressional interval transit time in formation matrix
	Δt_f Sonic compressional interval transit time in formation fluid in zone of investigation
	Δt_{sh} Sonic compressional interval transit time in shale
	B_{cp} Sonic compaction correction factor
	ϕ Formation porosity
Given:	Borehole fluid is fresh mud.

$\Delta t_c = 90\ \mu s/ft$

$\Delta t_{ma} = 55.6\ \mu s/ft$ (quartz)

$\Delta t_f = 189\ \mu s/ft$

(Estimated. This is a reasonable value for pore fluids near the wellbore when fresh muds are used.)

$\Delta t_{sh} = 110\ \mu s/ft$

Find: Empirical ϕ and time-average ϕ

Procedure: To determine the empirical ϕ, enter the chart on the Interval Transit Time axis at $\Delta t_c = 90\ \mu s/ft$. Project vertically to the empirical Quartz curve, then horizontally to the Porosity axis, there estimating ϕ to be **26.7%**.

To determine the time-average ϕ, calculate B_{cp}, noting that B_{cp} is defined as follows:

$B_{cp} = 1$ for limestones and dolomites

$B_{cp} = 1$ for sandstones where $\Delta t_{sh} \leq 100\ \mu s/ft$ in adjacent shales

$B_{cp} = \Delta t_{sh}/(100\ \mu s/ft)$ for sandstones where $\Delta t_{sh} > 100\ \mu s/ft$ in adjacent shales

So,

$$B_{cp} = \frac{110\ \mu s/ft}{100\ \mu s/ft} = 1.1$$

Enter the chart on the Interval Transit Time axis at $\Delta t_c = 90\ \mu s/ft$. Project vertically to the $B_{cp} = 1.1$ curve, then horizontally to the Porosity axis, there estimating ϕ to be 23.5%.

Answer: Empirical $\phi = 26.7\%$; time-average $\phi = 23.5\%$.

Notes: You can calculate the time-average ϕ in decimal form from the following equation:

$$\phi = \frac{\Delta t_c - \Delta t_{ma}}{\Delta t_f - \Delta t_{ma}} \cdot \frac{1}{B_{cp}}$$

Porosity Determination
Sonic vs. Porosity

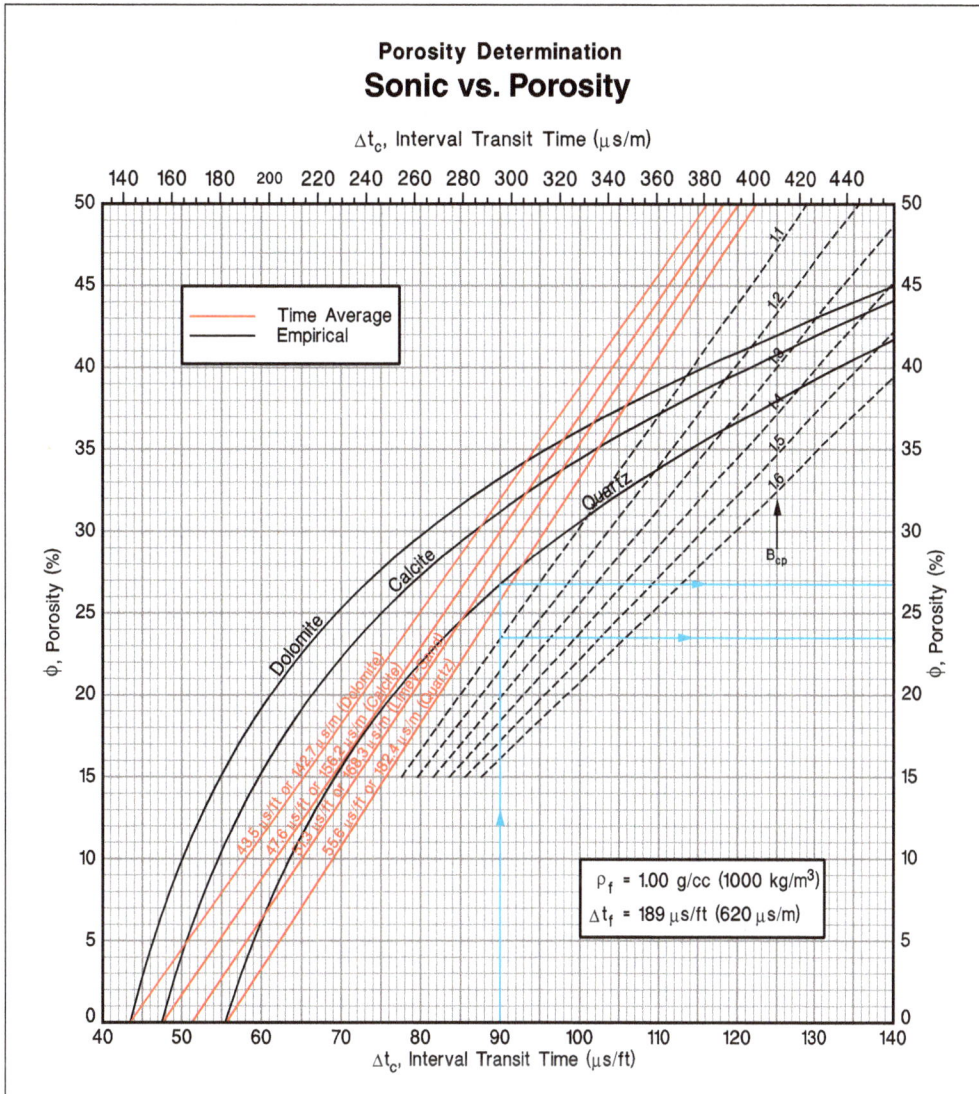

Fig. 3.9—POR-11 Sonic vs. porosity (Krief et al. 1989/ Halliburton Energy Services 1994).

POR-18 (Saturation Estimation In Gas-Bearing Zones)

Applications: Estimation of porosity and flushed zone water saturation in gas-bearing zones

Nomenclature:
T Formation temperature
P Formation pressure
ϕ_N Formation porosity from neutron measurement
ϕ_D Formation porosity from density measurement
ϕ_{DN} Formation porosity from combined neutron and density data
S_{XD} Flushed zone water saturation

Given: $T = 265°F$

$P = 10,300 \text{ psi}$

$\phi_N = 12\%$

$\phi_D = 32\%$

Find: ϕ_{DN} and S_{XD}

ρ_g Values	Temperature	Pressure	ρ_f
ρ_g = 0.1339 g/cc	125°F (51.67°C)	2,500 psi (17.25 Mpa)	0.996 g/cc
ρ_g = 0.2508 g/cc	275°F (135°C)	10,000 psi (68.95 Mpa)	0.960 g/cc

Table 3.1—Assumptions for determination of ρ_g values

Procedure: Since $T = 265°F \approx 275°F$ and $P = 10,300 \text{ psi} \approx 10,000$ psi, use the red curves in the chart. From $\phi_N = 12\%$ on the Neutron Porosity axis, project vertically into the chart. From $\phi_D = 32\%$ on the Density Porosity axis, project horizontally into the chart. The point of intersection of the two projections lies between the $\phi_{DN} = 20\%$ and $\phi_{DN} = 25\%$ curves (red curves with north-west-southeast orientation) and between the $S_{XD} = 20\%$ and $S_{XD} = 40\%$ curves (red curves with southwest-northeast orientation). Interpolation between appropriate curve pairs yields $\phi_{DN} = 23\%$ and $S_{XD} = 28\%$.

Answer: $\phi_{DN} = 23\%$ and $S_{XD} = 28\%$.

Notes: In constructing this chart, it was assumed that methane, ethane, and propane comprised 97.46% of the gas; i-butane, 0.96% ; and the following miscellaneous components:

Miscellaneous Components	Dry Gas (Mol. %)
Nitrogen (N_2)	0.11
Carbon dioxide (CO_2)	0.30
n-Butane (C_4H_{10})	0.40
i-Pentane (C_5H_{12})	0.31
n-Pentane (C_5H_{12})	0.11
Hexanes (C_6H_{14})	0.15
Heptanes Plus (C_7H_{16}, C_8H_{18}, C_9H_{20}, and $C_{10}H_{22}$)	0.5+0.5+0.5+0.5 = 0.20

Table 3.2—Miscellaneous component percentages in dry gas.

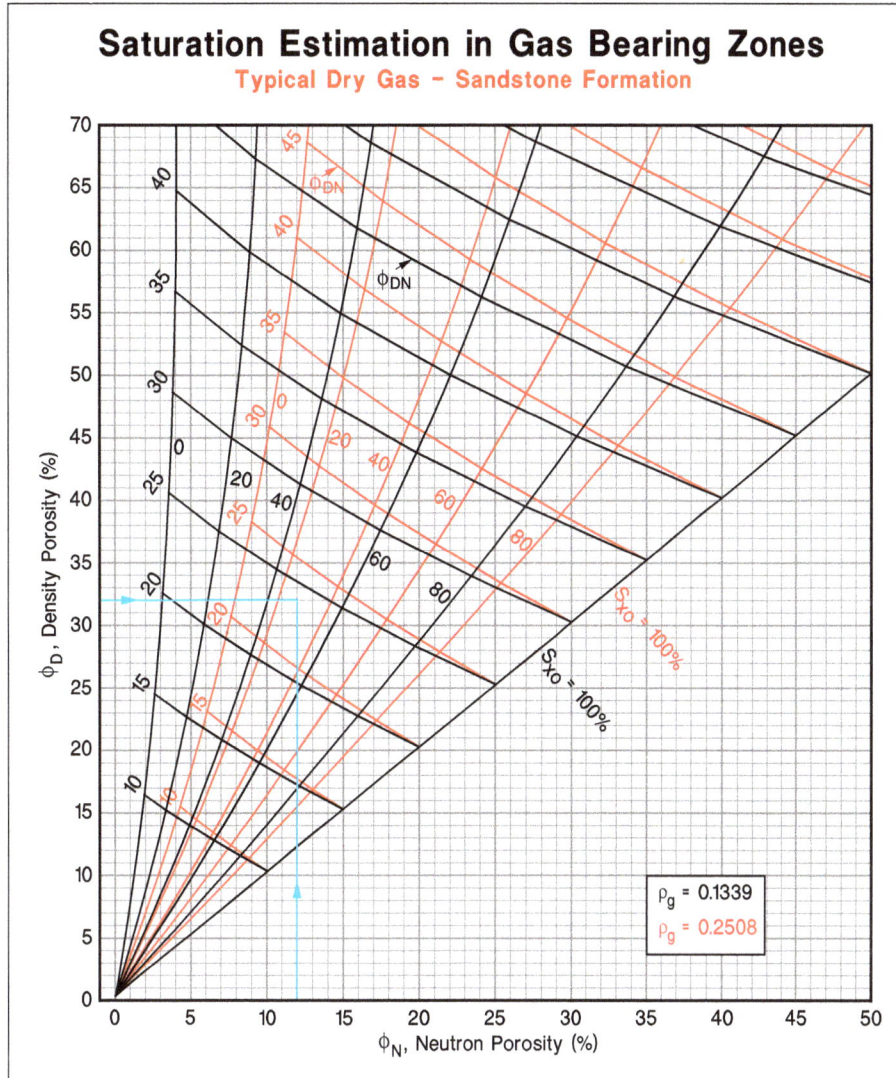

Fig. 3.10—POR-18 Saturation estimation in gas-bearing zones (Halliburton Energy Services 1994).

CBL-3 (Free-Pipe Amplitude and Attenuation)

Applications: Determination of free-pipe acoustic amplitude and attenuation from casing size

Nomenclature: A Amplitude of acoustic measurement

OD_{csg} Casing outside diameter

α Acoustic attenuation rate

Example 1

Given: $OD_{csg} = 5.5$ in.

Find: A

Procedure: Enter the chart at $OD_{csg} = 5.5$ in on the Casing Size axis. Project vertically until reaching the Amplitude curve, then horizontally to the left until intersecting the E1 Amplitude axis, there estimating A to be approximately 72 mV.

Answer: $A = 72$ mV

Notes: You can calculate A from

$$A = 201.54 \cdot OD_{csg}^{-0.6044}$$

where A is in mV and OD_{csg} is in in.

Example 2

Given: $OD_{csg} = 8.5$ in.

Find: α

Procedure: Enter the chart at $OD = 8.5$ in. on the Casing Size axis. Project vertically until reaching the Attenuation Rate curve, then horizontally to the right until intersecting the E1 Attenuation axis, there estimating α to be approximately 0.30 dB/ft.

Answer: $\alpha = 0.30$ dB/ft

Notes: You can calculate α from

$$\alpha = 0.0189 \cdot (1.0512)^{A}$$

where α is in dB/ft and A is in mV.

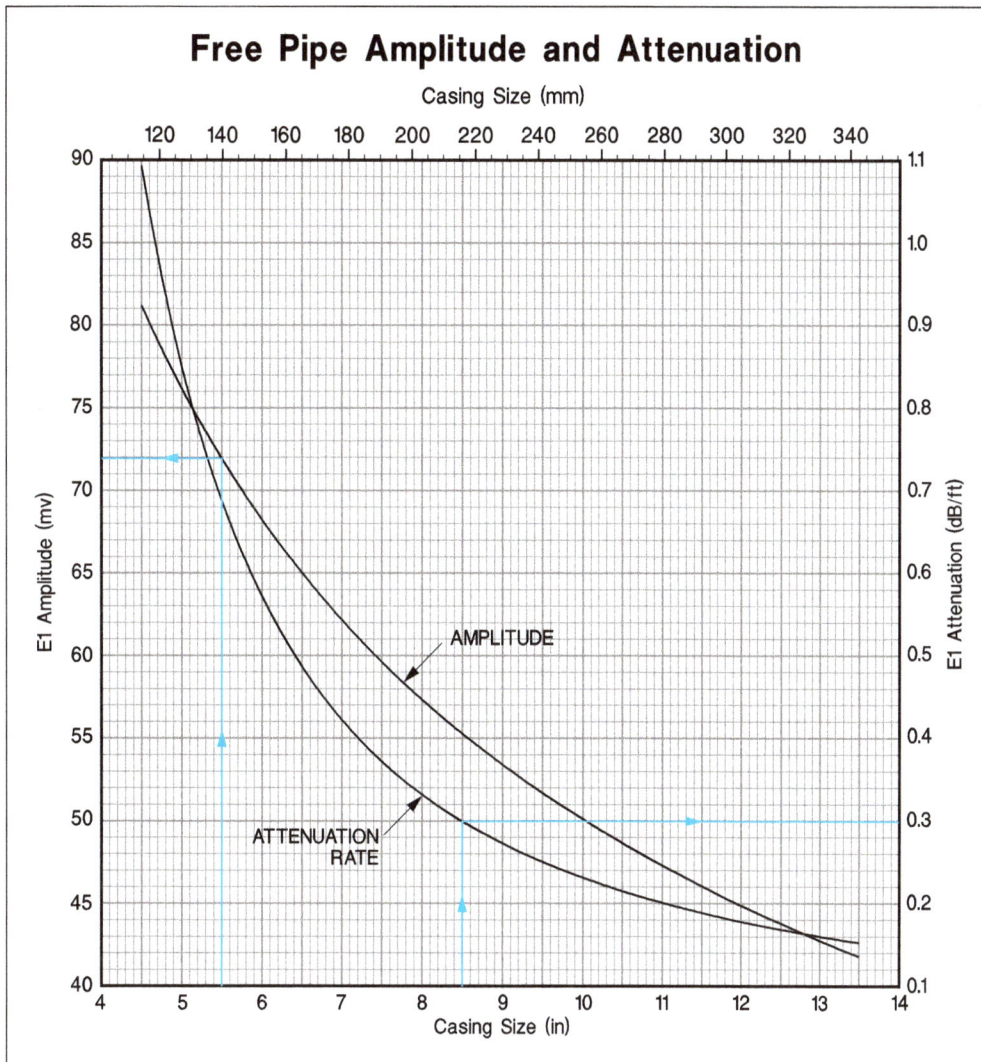

Fig. 3.11—CBL-3 Free pipe amplitude and attenuation (Halliburton Energy Services 1994).

CBL-4 (CBL Borehole Fluid Attenuation Effects; see Fig. 3.12)

Applications:	Determination of borehole fluid effects on acoustic attenuation
Nomenclature:	A Amplitude of acoustic measurement
	A_{H_2O} Amplitude of acoustic measurement in water
	ρ_t Fluid density

Example 1

Given: $\rho_t = 12.5$ lbm/gal (completion fluid)

Find: A/A_{H_2O}

Procedure: Enter the chart at $\rho_t = 12.5$ lbm/gal on the Fluid Weight axis. Project vertically until reaching the Completion Fluids curve, then horizontally until intersecting the Amplitude Ratio axis, and there estimating A/A_{H_2O} to be approximately 1.60 (i.e., the acoustic amplitude in the completion fluid is approximately 1.60 times the acoustic amplitude in water).

Answer: $A/A_{H_2O} = 1.60$.

     If 5.5-in. OD casing were in this well and were filled with water, the free-pipe amplitude would be 72 mV (see chart CBL-3 in Fig. 3.11). However, because 12.5-lbm/gal completion fluid is in the well, the free-pipe amplitude would be 72 mV 1.6=115 mV.

Notes: You can calculate A/A_{H_2O} from

$$A/A_{H_2O} = \left(0.128 \cdot \rho_f\right) - 0.00482,$$

where ρ_f is in lbm/gal.

     Also, if you know A and have determined A/A_{H_2O}, then you can calculate A_{H_2O} by dividing A by A/A_{H_2O}.

Example 2

Given: $\rho_f = 15.0$ lbm/gal (mud)

Find: A/A_{H_2O}

Procedure: Enter the chart at $\rho_f = 15.0$ lbm/gal on the Fluid Weight axis. Project vertically until reaching the Mud curve, then horizontally until intersecting the Amplitude Ratio axis, and there estimating A/A_{H_2O} to be approximately 1.15 (i.e., the acoustic amplitude in the mud is approximately 1.15 times the acoustic amplitude in water).

Answer: $A/A_{H_2O} = 1.15$

Notes: You can calculate A/A_{H_2O} from

$$\frac{A}{A_{H_2O}} = \left(0.5121 \cdot \rho_f\right) - 0.4875,$$

where ρ_f is in lbm/gal.

     Also, if you know A and have determined A/A_{H_2O}, then you can calculate A_{H_2O} by dividing A by A/A_{H_2O}.

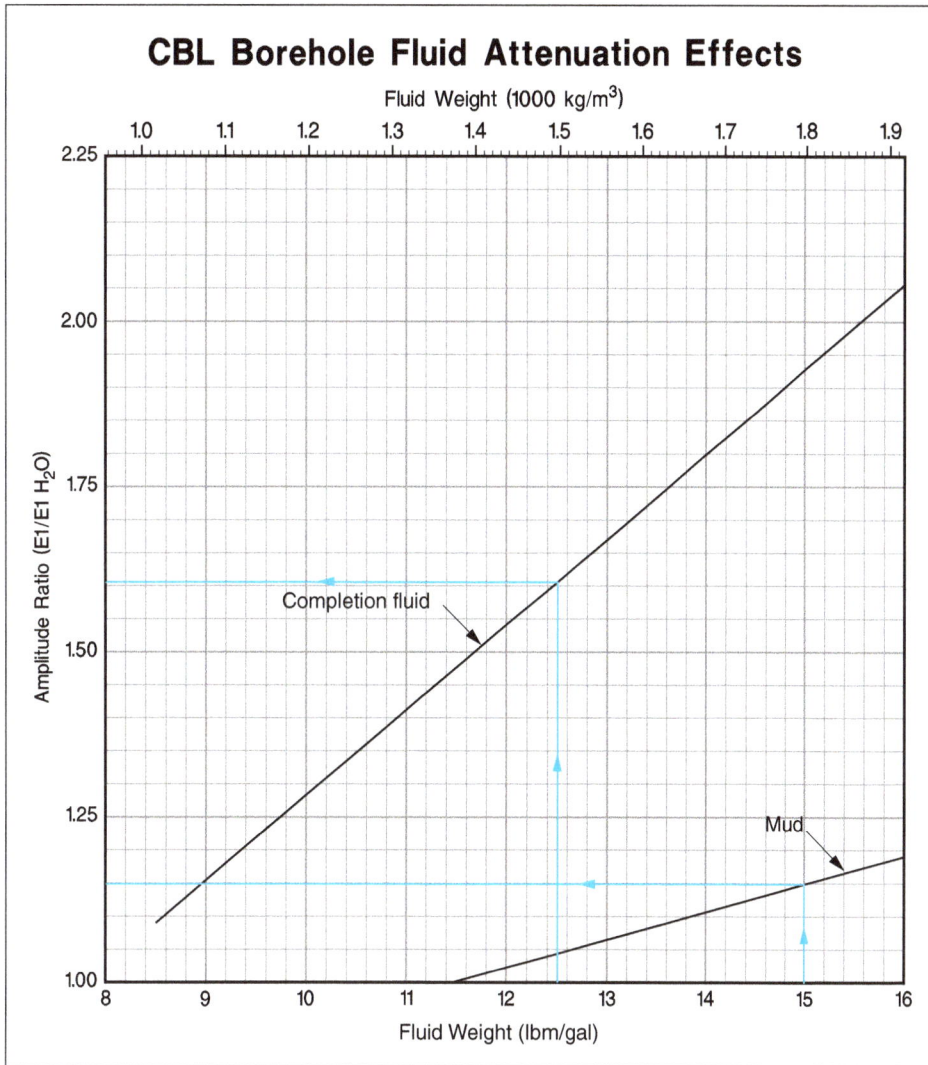

Fig. 3.12—CBL-4 borehole fluid attenuation effects (Halliburton Energy Services 1994).

Chapter 4

Production Engineering

4.1 Inflow/Outflow Performance

Gas Well Performance (Rawlins and Schellhardt 1935)

$$\frac{q_g}{q_{g,\max}} = \left[1 - \left(\frac{p_{wf}}{\bar{p}_R} \right)^2 \right]^n$$

$$\frac{q_g}{q_{g,\max}} = \left[1 - \frac{p_p(p_{wf})}{p_p(\bar{p}_R)} \right]^n$$

n	Deliverability exponent
q_g	Gas flow rate, Mscf/D
$q_{g,\max}$	Absolute open flow (AOF), maximum gas flow rate, Mscf/D
$\bar{p}_R$	Average reservoir pressure, psia
p_{wf}	Bottomhole pressure, psia
p_p	Real-gas pseudopressure, psia^2/cp

Gas Well Performance (Houpeurt 1959)

$$\bar{p}_R^2 - p_{wf}^2 = aq_g + bq_g^2$$

$$a = \frac{1{,}422\,\mu z T \left(\ln \dfrac{r_e}{r_w} - \dfrac{3}{4} + s \right)}{kh}$$

$$b = \frac{1{,}422\,\mu z T}{kh} D$$

$$p_p\left(\overline{p}_R\right) - p_p\left(p_{wf}\right) = aq_g + bq_g^2$$

$$a = \frac{1{,}422T\left(\ln\dfrac{r_e}{r_w} - \dfrac{3}{4} + s\right)}{kh}$$

$$b = \frac{1{,}422T}{kh}D$$

$$D = \frac{2.715 \times 10^{-15}\,\beta kM p_{sc}}{h\mu_g r_w T_{sc}}$$

$$\beta = 1.88 \times 10^{10}\,k^{-1.47}\phi^{-0.53}$$

q_g	Gas flow rate, Mscf/D
$\overline{p}_R$	Average reservoir pressure, psia
p_{wf}	Bottomhole pressure, psia
p_p	Real-gas pseudopressure, psia^2/cp
T	Temperature, °R
z	Gas compressibility factor, dimensionless
μ	Viscosity, cp
μ_g	Gas viscosity, cp
r_e	External drainage radius, ft
r_w	Wellbore radius, ft
k	Permeability, md
h	Formation thickness, ft
s	Skin factor, dimensionless
a	Laminar flow coefficient, psia^2/Mscf/D
b	Turbulence coefficient, psia^2/(Mscf/D)^2
D	Non-Darcy flow coefficient, D/Mscf
β	Turbulence factor, ft^{-1}
ϕ	Porosity, fraction
M	Molecular weight, lbm/lbm-mol
p_{sc}	Standard pressure, psia
T_{sc}	Standard temperature, °R

Oil Well Inflow-Performance Relationships

Vogel's Method

$$\frac{q_o}{q_{o,\max}} = 1 - 0.2\left(\frac{p_{wf}}{\overline{p}_R}\right) - 0.8\left(\frac{p_{wf}}{\overline{p}_R}\right)^2$$

$$J = \frac{q_o}{\overline{p}_R - p_{wf}}, \qquad \overline{p}_R \ge p_{wf} \ge p_b$$

$$J = \frac{q_o}{\overline{p}_R - p_b + \dfrac{p_b}{1.8}\left[1.0 - 0.2\dfrac{p_{wf}}{p_b} - 0.8\left(\dfrac{p_{wf}}{p_b}\right)^2\right]}, \qquad \overline{p}_R \ge p_b \ge p_{wf}$$

Wiggins' Method

$$\frac{q_o}{q_{o,\max}} = 1 - 0.52\left(\frac{p_{wf}}{\overline{p}_R}\right) - 0.48\left(\frac{p_{wf}}{\overline{p}_R}\right)^2$$

$$\frac{q_w}{q_{w,\max}} = 1 - 0.72\left(\frac{p_{wf}}{\overline{p}_R}\right) - 0.28\left(\frac{p_{wf}}{\overline{p}_R}\right)^2$$

Standing's Method

$$\frac{q_o}{q_{o,\max}} = \left(1 - \frac{p_{wf}}{\overline{p}_R}\right)\left[1 + 0.8\left(\frac{p_{wf}}{\overline{p}_R}\right)\right]$$

The Klins-Clark Method

$$\frac{q_o}{q_{o,\max}} = 1 - 0.295\left(\frac{p_{wf}}{\overline{p}_R}\right) - 0.705\left(\frac{p_{wf}}{\overline{p}_R}\right)^d$$

$$d = \left[0.28 + 0.72\left(\frac{\overline{p}_R}{p_b}\right)\right](1.24 + 0.001p_b)$$

Fetkovich's Method

$$q_o = \frac{0.00708kh}{\left(\ln\frac{r_e}{r_w} - 0.75 + s\right)}\int_{p_{wf}}^{\overline{p}_R} f(p)\,\mathrm{d}p$$

$$f(p) = \frac{k_{ro}}{\mu_o B_o}$$

d	Decline factor
k	Absolute permeability, md
h	Pay thickness, ft
q_o	Oil flow rate, STB/D
$q_{o,\max}$	Maximum oil flow rate, STB/D
q_w	Water flow rate, STB/D
$q_{w,\max}$	Maximum water flow rate, STB/D
$\overline{p}_R$	Average reservoir pressure, psia
p_{wf}	Bottomhole pressure, psia
p_b	Bubblepoint pressure, psia
k_{ro}	Oil relative permeability, md
B_o	Oil formation volume factor, bbl/STB
μ_o	Oil viscosity, cp
J	Productivity index, STB/D/psia
r_w	Wellbore radius, ft
r_e	External boundary radius, ft

Pressure Drop in Gas Wells

$$q = 200 \left[\frac{Sd^5 \left(p_{wf}^2 - e^S p_{wh}^2 \right)}{\gamma_g TzLf_M \left(e^S - 1 \right)} \right]$$

$$f_M = \left\{ 2\log\left[3.71/(\varepsilon/d) \right] \right\}^{-2}$$

$$S = \frac{0.0375\gamma_g L}{Tz}$$

q	Flow rate, Mscf/D
d	Pipe diameter, in.
p_{wf}	Bottomhole pressure, psia
p_{wh}	Wellhead pressure, psia
γ_g	Gas-specific gravity, dimensionless
L	Length of pipe, psia
T	Temperature, °R
z	Gas compressibility factor, dimensionless
f_M	Moody friction factor, dimensionless
ε	Absolute pipe roughness, in.

Skin Effect

Pressure Drop Caused by Skin

$$\Delta p_s = 0.87ms$$

Δp_s	Pressure drop caused by skin, psi
m	Slope from Homer's Plot, psi/cycle
s	Skin

Flow Efficiency

$$FE = \frac{\left(P_r - P_{wf} - \Delta P_{skin} \right)}{\left(P_r - P_{wf} \right)}$$

FE	Flow efficiency
P_r	Reservoir pressure, psi
P_{wf}	Producing pressure, psi
ΔP_{skin}	Pressure drop caused by skin, psi

Gas Velocity

$$V = \frac{60QzT}{D^2 P}$$

V	Gas velocity, ft/sec
Q	Flow rate, MMcf/D
z	Gas compressibility
T	Temperature, °R
D	Tubular internal diameter, in.
P	Pressure, psia

4.2 Impact of Length and Force Changes to the Tubing String

Length Change Calculations

$$\Delta L_{\text{stretch}} = \Delta L_a + \Delta L_p + \Delta L_b + \Delta L_t$$

$\Delta L_{\text{stretch}}$ Total stretch or elongation of the drillpipe, ft

ΔL_a Stretch caused by axial load, ft

$$\Delta L_a = \frac{F_T \times L}{A \times E} + \frac{\Delta F_T \times L}{2 \times A \times E}$$

F_T Pressure-area axial force, lbf

ΔF_T Change in pressure-area axial force over the component length, lbf

A Cross-sectional area of the component, in.^{2}

E Young's modulus of the component material, psi

L Length, ft

ΔL_p Stretch resulting from the pressure effect (ballooning), ft

$$\Delta L_p = \frac{-\upsilon \times L_p}{E \times (R^2 - 1)} \times \left[(\rho_s - R^2 \times \rho_a) \times L + 2 \times (P_s - R^2 \times P_a) \right]$$

L_p Length of the workstring-component element, ft

R Ratio of the component's OD to the ID

υ Poisson's ratio of the component material

ρ_s Mud density inside the workstring component, ppg

ρ_a Mud density in the annulus at the depth of the workstring component, ppg

P_s Surface pressure on the drillstring side, psi

P_a Surface pressure in the annulus side, psi

ΔL_b Stretch resulting from buckling, ft

$$\Delta L_b = \frac{-r^2}{4EI\omega}(F_2 - F_p)(0.3771F_2 - 0.3668F_p) \qquad \text{for } 2.8F_p > F_2 > F_p$$

$$\Delta L_b = \frac{-r^2}{8EI\omega}(F_2^2 - F_1^2) \qquad \text{for } F > 2.8F_p$$

F_1, F_2 Buckling force, lbf

F_p Paslay buckling force

ω Distributed buoyed weight of the casing

EI Pipe-bending stiffness

r Radial annular clearance

ΔL_t Stretch resulting from temperature, ft

$$\Delta L_t = a_t \left(\Delta t_0 + \frac{\Delta t}{\Delta z} \frac{L^2}{2} \right)$$

Δt_0 Initial average temperature change, °F

Δt Average temperature change, °F

Δz Change in measured depth, ft

L Measured calculation interval, ft

a_t Coefficient of thermal expansion defined as the fractional increase in length per unit rise in temperature, in./in./°F, = 6.9×10^{-6} for steel, 10.3×10^{-6} for aluminum, and 4.9×10^{-6} for titanium

4.3 Tubing Design

Maximum Hydrostatic Test Pressure

$$p_h = \frac{2 \times \left(0.8 \times \sigma_y \times t\right)}{d_o}$$

Internal Yield Pressure Rating for Tubing

$$p_{yi} = \frac{0.875 \times 2 \times \sigma_y \times t}{d_o}$$

p_h	Hydrostatic test pressure, psi
p_{yi}	Internal yield pressure, psi
σ_y	Minimum yield strength of pipe, psi
t	Tube thickness, in.
d_o	Tubing OD, in.

Stretch in Tubing

$$\Delta L_t = \frac{12 \times L_p \times F}{E \times A_m}$$

ΔL_t	Total axial stretch or contraction, in.
L_p	Length of pipe, ft
F	Superimposed tension or compression axial load, lbf
E	Young's modulus of elasticity for steel, = 30 million psi
A_m	Cross section of metal area of pipe, in.^{2}

Tubing Hook Load in Air

$$F_a = L_p \times w_n$$

F_a	Tubing hook load in air, lbf
L_p	Length of pipe, ft
w_n	Weight per foot of tubing, lbm/ft

Tubing Hook Load in Fluid

$$F_f = F_a - F_b$$

F_f	Tubing hook load in fluid, lbf
F_a	Tubing hook load in air, lbf
F_b	Axial buoyancy load, lbf

Collapse Differential Pressure

$$\Delta p_c = \left(g_a - g_t\right) \times D_{tV}$$

Δp_c	Collapse differential, psi
g_a	Gradient in the annulus, psi/ft
g_t	Gradient in the tubing, psi/ft
D_{tV}	True vertical depth, ft

Burst Differential Pressure

$$\Delta p_b = \left(g_t - g_a\right) \times D_{tV} + p_{wh}$$

Δp_b Burst differential, psi
g_a Gradient in the annulus, psi/ft
g_t Gradient in the tubing, psi/ft
D_{tV} True vertical depth, ft
p_{wh} Wellhead pressure, psi

Length of Tubing

$$L_p = \frac{\left[\left(p_{wh} - p_{br}\right)/D_b\right]}{g_a - g_t}$$

L_p Length of pipe, ft
p_{wh} Wellhead pressure, psi
p_{br} Burst pressure rating, psi
D_b Design factor in burst, dimensionless
g_a Gradient in the annulus, psi/ft
g_t Gradient in the tubing, psi/ft

Effect of Pressure and Temperature Changes on Tubing Sealed in a Packer

Hooke's Law Effect

$$\Delta L = \frac{L \Delta F}{E A_s}$$

$$\Delta F = \left(A_p - A_i\right)\Delta P_i - \left(A_p - A_o\right)\Delta P_o$$

ΔL Change in tubing length resulting from Hooke's law effect, in.
L Length of tubing, in.
F Force acting on bottom of tubing, lb
E Modulus of elasticity, 30×10^6 psi for steel
A_s Cross-sectional area of tubing, in.^{2}
A_i Area based on ID of tubing, in.^{2}
A_o Area based on outside diameter of tubing, in.^{2}
A_p Area based on diameter of packer seal, in.^{2}
P_o Pressure at packer seal in annulus, psi
P_i Pressure at packer seal in tubing, psi

Buckling Effect

$$\Delta L = \frac{r^2 A_p^2 \left(\Delta P_i - \Delta P_o\right)^2}{8\,E\,I\,\left(w_s + w_i - w_o\right)}$$

$$I = \frac{\pi}{64}\left(D^4 - d^4\right)$$

ΔL Change in tubing length resulting from buckling, in.
r Radial clearance between tubing and casing, in.
w $= w_s + w_i - w_o$
w_s Weight of tubing, lbm/in.
w_i Weight of fluid contained inside tubing, lbm/in.

117

w_o Weight of annulus fluid displaced by bulk volume of tubing, lbm/in.

D Tubing OD, in.

d Tubing ID, in.

Ballooning Force

$$F_{Ba} = -0.6\left(\Delta p_{i,a} A_i - \Delta p_{o,a} A_o\right)$$

F_{Ba} Force created by ballooning, lbf

$\Delta p_{i,a}$ Average internal pressure change, psi

$\Delta p_{o,a}$ Average external pressure change, psi

A_i Tubing inner area, in.^{2}

A_o Tubing outer area, in.^{2}

Ballooning Effect

$$\Delta L = \frac{\mu L^2}{E}\frac{\Delta \rho_i - R^2 \Delta \rho_o - \dfrac{1+2\mu}{2\mu}\delta}{R^2 - 1} + \frac{2\mu L}{E}\left(\frac{\Delta p_i - R^2 \Delta p_o}{R^2 - 1}\right)$$

ΔL Change in length caused by ballooning, in.

μ Poisson's ration (0.3 for steel)

R Tubing OD/Tubing ID

$\Delta \rho_i$ Change in density of fluid inside tubing, lbf/in.^{3}

$\Delta \rho_o$ Change in density of fluid outside tubing, lbf/in.^{3}

Δp_i Change in surface pressure inside tubing, psi

Δp_o Change in surface pressure outside tubing, psi

δ Pressure drop in tubing because of flow, psi/in. (usually considered 0)

Temperature Effect

$$\Delta L = LC\Delta T$$

ΔL Change in length resulting from temperature, in.

L Length of tubing string, in.

C Coefficient of expansion of steel per °F, $= 6.9 \times 10^{-6}$

ΔT Temperature change, °F

$$\Delta F = 207 A_s \Delta T$$

ΔF Change in force acting on tubing, lbf

A_s Cross-sectional area of tubing, in.^{2}

ΔT Temperature change, °F

TABLE 3.12—MINIMUM PERFORMANCE PROPERTIES OF TUBING (Reproduced courtesy of the American Petroleum Institute from API BULL. 5C2[6])

1	2	3	4	5	6	7	8	9	10	11	12	13	14	15	16	17	18	19	20	21	22	23	24
	Nominal Weight							Coupling Outside Diameter / Integral Joint						Internal Yield Pressure						Joint Yield Strength (F_j)			
	Threaded and Coupled		Integral Joint					Threaded and Coupled	Upset		Integral Joint					Threaded and Coupled Upset					Threaded and Coupled Upset		
Size OD, D (in.)	Nonupset (lbm/ft)	Upset (lbm/ft)	Integral Joint (lbm/ft)	Grade	Wall Thickness, t (in.)	d (in.)	Drift Diameter (in.)	Nonupset W (in.)	Regular W (in.)	Special Clearance W_c (in.)	Drift Diameter (in.)	OD of Box (in.)	Collapse Resistance	Plain End (psi)	Nonupset (psi)	Regular (psi)	Special Clearance (psi)	Integral Joint (psi)	Pipe Body Yield (lbf)	Nonupset (lbf)	Regular (lbf)	Special Clearance (lbf)	Integral Joint (lbf)
1.050	1.14	1.20	—	H-40	.113	.824	.730	1.313	1.660	—	—	—	7,680	7,530	7,530	7,530	—	—	13,320	6,320	13,320	—	—
1.050	—	1.54	—	H-40	.154	.742	.648	—	1.660	—	—	—	10,010	10,270	—	9,420	—	—	17,320	—	17,320	—	—
1.050	1.14	1.20	—	J-55	.113	.824	.730	1.313	1.660	—	—	—	10,560	10,360	10,360	10,360	—	—	18,320	8,690	18,320	—	—
1.050	—	1.54	—	J-55	.154	.742	.648	—	1.660	—	—	—	13,770	14,120	—	12,950	—	—	23,820	—	23,820	—	—
1.050	1.14	1.20	—	L-80	.113	.824	.730	1.313	1.660	—	—	—	15,370	15,070	15,070	15,070	—	—	26,640	12,640	26,640	—	—
1.050	—	1.54	—	L-80	.154	.742	.648	—	1.660	—	—	—	20,020	20,530	—	18,840	—	—	34,640	—	34,640	—	—
1.050	1.14	1.20	—	N-80	.113	.824	.730	1.313	1.660	—	—	—	15,370	15,070	15,070	15,070	—	—	26,640	12,640	26,640	—	—
1.050	—	1.54	—	N-80	.154	.742	.648	—	1.660	—	—	—	20,020	20,530	—	18,840	—	—	34,640	—	34,640	—	—
1.050	1.14	1.20	—	C-90	.113	.824	.730	1.313	1.660	—	—	—	17,290	16,950	16,950	16,950	—	—	29,970	14,220	29,970	—	—
1.050	—	1.54	—	C-90	.154	.742	.648	—	1.660	—	—	—	22,530	23,100	—	21,200	—	—	38,970	—	38,970	—	—
1.050	1.14	1.20	—	T-95	.113	.824	.730	1.313	1.660	—	—	—	18,250	17,890	17,890	17,890	—	—	31,640	15,010	31,640	—	—
1.050	—	1.54	—	T-95	.154	.742	.648	—	1.660	—	—	—	23,780	24,389	—	22,380	—	—	41,140	—	41,140	—	—
1.050	—	1.54	—	P-110	.154	.742	.648	—	1.660	—	—	—	27,530	28,230	—	25,910	—	—	47,630	—	47,630	—	—
1.315	1.70	1.80	1.72	H-40	.133	1.049	.955	1.660	1.900	—	.955	1.550	7,270	7,080	7,080	7,080	—	7,080	19,760	10,920	19,760	—	15,940
1.315	—	2.24	—	H-40	.179	.957	.863	—	1.900	—	—	—	9,410	9,530	—	9,530	—	—	25,560	—	25,560	—	—
1.315	1.70	1.80	1.72	J-55	.133	1.049	.955	1.660	1.900	—	.955	1.550	10,000	9,730	9,730	9,730	—	9,730	27,170	15,020	27,170	—	21,910
1.315	—	2.24	—	J-55	.179	.957	.863	—	1.900	—	—	—	12,940	13,100	—	13,100	—	—	35,150	—	35,150	—	—
1.315	1.70	1.80	1.72	L-80	.133	1.049	.955	1.660	1.900	—	.955	1.550	14,550	14,160	14,160	14,160	—	14,160	39,520	21,840	39,520	—	31,870
1.315	—	2.24	—	L-80	.179	.957	.863	—	1.900	—	—	—	18,810	19,060	—	19,060	—	—	51,120	—	51,120	—	—
1.315	1.70	1.80	1.72	N-80	.133	1.049	.955	1.660	1.900	—	.955	1.550	14,550	14,160	14,160	14,160	—	14,160	39,520	21,840	39,520	—	31,870
1.315	—	2.24	—	N-80	.179	.957	.863	—	1.900	—	—	—	18,810	19,060	—	19,060	—	—	51,120	—	51,120	—	—
1.315	1.70	1.80	1.72	C-90	.133	1.049	.955	1.660	1.900	—	.955	1.550	16,360	15,930	15,930	15,930	—	15,930	44,460	24,570	44,460	—	35,860
1.315	—	2.24	—	C-90	.179	.957	.863	—	1.900	—	—	—	21,170	21,440	—	21,440	—	—	57,510	—	57,510	—	—
1.315	1.70	1.80	1.72	T-95	.133	1.049	.955	1.660	1.900	—	.955	1.550	17,270	16,810	16,810	16,810	—	16,810	46,930	25,940	46,930	—	37,850
1.315	—	2.24	—	T-95	.179	.957	.863	—	1.900	—	—	—	22,340	22,630	—	22,630	—	—	60,710	—	60,710	—	—
1.315	—	2.24	—	P-110	.179	.957	.863	—	1.900	—	—	—	25,870	26,200	—	26,200	—	—	70,290	—	70,290	—	—
1.660	—	—	2.10	H-40	.125	1.410	1.316	—	—	—	1.316	1.880	5,570	5,270	—	—	—	5,270	24,120	—	—	—	22,230
1.660	2.30	2.40	2.33	H-40	.140	1.380	1.286	2.054	2.200	—	1.286	1.880	6,180	5,900	5,900	5,900	—	5,810	26,760	15,480	26,760	—	22,230
1.660	—	3.07	—	H-40	.191	1.278	1.184	—	2.200	—	—	—	8,150	8,050	—	8,050	—	—	35,240	—	35,240	—	—
1.660	—	—	2.10	J-55	.125	1.410	1.316	—	—	—	1.316	1.880	7,660	7,250	—	—	—	7,250	33,170	—	—	—	30,560
1.660	2.30	2.40	2.33	J-55	.140	1.380	1.286	2.054	2.200	—	1.286	1.880	8,490	8,120	8,120	8,120	—	7,990	36,800	21,290	36,800	—	30,560
1.660	—	3.07	—	J-55	.191	1.278	1.184	—	2.200	—	—	—	11,200	11,070	—	11,070	—	—	48,460	—	48,460	—	—
1.660	2.30	2.40	2.33	L-80	.140	1.380	1.286	2.054	2.200	—	1.286	1.880	12,360	11,810	11,810	11,810	—	11,620	53,520	30,960	53,520	—	44,460
1.660	—	3.07	—	L-80	.191	1.278	1.184	—	2.200	—	—	—	16,290	16,110	—	16,110	—	—	70,480	—	70,480	—	—
1.660	2.30	2.40	2.33	N-80	.140	1.380	1.286	2.054	2.200	—	1.286	1.880	12,360	11,810	11,810	11,810	—	11,620	53,520	30,960	53,520	—	44,460
1.660	—	3.07	—	N-80	.191	1.278	1.184	—	2.200	—	—	—	16,290	16,110	—	16,110	—	—	70,480	—	70,480	—	—
1.660	2.30	2.40	2.33	C-90	.140	1.380	1.286	2.054	2.200	—	1.286	1.880	13,900	13,280	13,280	13,280	—	13,070	60,210	34,830	60,210	—	50,010
1.660	—	3.07	—	C-90	.191	1.278	1.184	—	2.200	—	—	—	18,330	18,120	—	18,120	—	—	79,290	—	79,290	—	—
1.660	2.30	2.40	2.33	T-95	.140	1.380	1.286	2.054	2.200	—	1.286	1.880	14,670	14,020	14,020	14,020	—	13,800	63,560	36,770	63,560	—	52,790
1.660	—	3.07	—	T-95	.191	1.278	1.184	—	2.200	—	—	—	19,350	19,130	—	19,130	—	—	83,700	—	83,700	—	—
1.660	—	3.07	—	P-110	.191	1.278	1.184	—	2.200	—	—	—	22,400	22,150	—	22,150	—	—	96,910	—	96,910	—	—
1.900	—	—	2.40	H-40	.125	1.650	1.556	—	—	—	1.556	2.110	4,920	4,610	—	—	—	4,610	27,880	—	—	—	26,940
1.900	2.75	2.90	2.76	H-40	.145	1.610	1.516	2.200	2.500	—	1.516	2.110	5,640	5,340	5,340	5,340	—	5,140	31,960	19,040	31,960	—	26,940
1.900	—	3.73	—	H-40	.200	1.500	1.406	—	2.500	—	—	—	7,530	7,370	—	7,370	—	—	42,720	—	42,720	—	—
1.900	—	—	2.40	J-55	.125	1.650	1.556	—	—	—	1.556	2.110	6,640	6,330	—	—	—	6,330	38,340	—	—	—	37,040
1.900	2.75	2.90	2.76	J-55	.145	1.610	1.516	2.200	2.500	—	1.516	2.110	7,750	7,350	7,350	7,350	—	7,060	43,950	26,180	43,950	—	37,040
1.900	—	3.73	—	J-55	.200	1.500	1.406	—	2.500	—	—	—	10,360	10,130	—	10,130	—	—	58,740	—	58,740	—	—
1.900	2.75	2.90	2.76	L-80	.145	1.610	1.516	2.200	2.500	—	1.516	2.110	11,280	10,680	10,680	10,680	—	10,270	63,920	38,080	63,920	—	53,880
1.900	—	3.73	—	L-80	.200	1.500	1.406	—	2.500	—	—	—	15,070	14,740	—	14,740	—	—	85,440	—	85,440	—	—
1.900	4.42	—	—	L-80	.250	1.400	1.306	—	—	—	—	—	18,280	18,420	—	—	—	—	103,680	—	—	—	—
1.900	5.15	—	—	L-80	.300	1.300	1.206	—	—	—	—	—	21,270	22,110	—	—	—	—	120,640	—	—	—	—
1.900	2.75	2.90	2.76	N-80	.145	1.610	1.516	2.200	2.500	—	1.516	2.110	11,280	10,680	10,680	10,680	—	10,270	63,920	38,080	63,920	—	53,880

Table 3.12 shows minimum tubing performance properties.

TABLE 3.12—MINIMUM PERFORMANCE PROPERTIES OF TUBING (continued)

1	2	3	4	5	6	7	8	9	10	11	12	13	14	15	16	17	18	19	20	21	22	23	24
	Nominal Weight							Threaded and Coupled — Coupling Outside Diameter			Integral Joint			Internal Yield Pressure		Threaded and Coupled				Joint Yield Strength (F)		Threaded and Coupled	
	Threaded and Coupled								Upset							Upset						Upset	
Size OD, D (in.)	Non-upset (lbm/ft)	Upset (lbm/ft)	Integral Joint (lbm/ft)	Grade	Wall Thickness t (in.)	d (in.)	Drift Diameter (in.)	Non-upset W (in.)	Regular W (in.)	Special Clearance Wc (in.)	Drift Diameter (in.)	OD of Box (in.)	Collapse Resistance	Plain End	Nonupset (psi)	Regular (psi)	Special Clearance (psi)	Integral Joint (lbf)	Pipe Body Yield (lbf)	Nonupset (lbf)	Regular (lbf)	Special Clearance (lbf)	Integral Joint (lbf)
1.900	2.75	—	—	N-80	.200	1.500	1.406	—	2.500	—	—	—	15,070	14,740	—	14,740	—	—	85,440	—	85,440	—	—
1.900	—	2.90	2.76	C-90	.145	1.610	1.516	2.000	2.500	—	1.516	2.110	12,620	12,020	12,020	12,020	—	11,560	71,910	42,840	71,910	—	60,610
1.900	4.42	—	—	C-90	.200	1.500	1.406	—	2.500	—	—	—	16,950	16,580	—	16,580	—	—	96,120	—	96,120	—	—
1.900	5.15	—	—	C-90	.250	1.400	1.306	—	—	—	—	—	20,570	20,720	—	—	—	—	116,640	—	—	—	—
1.900	—	—	—	C-90	.300	1.300	1.206	—	—	—	—	—	23,930	24,870	—	—	—	—	135,720	—	—	—	—
1.900	2.75	2.90	2.76	T-95	.145	1.610	1.516	2.200	2.500	—	1.516	2.110	13,190	12,690	12,690	12,690	—	12,200	75,910	45,220	75,910	—	63,980
1.900	—	3.73	—	T-95	.200	1.500	1.406	—	2.500	—	—	—	17,890	17,500	—	17,500	—	—	101,460	—	101,460	—	—
1.900	4.42	—	—	T-95	.250	1.400	1.306	—	—	—	—	—	21,710	21,880	—	—	—	—	123,120	—	—	—	—
1.900	5.15	—	—	T-95	.300	1.300	1.206	—	—	—	—	—	25,260	26,250	—	—	—	—	143,260	—	—	—	—
1.900	—	3.73	—	P-110	.200	1.500	1.406	—	2.500	—	—	—	20,720	20,260	—	20,260	—	—	117,480	—	117,480	—	—
1.900	—	—	3.25	H-40	.156	1.751	1.657	—	—	—	1.657	2.325	5,590	5,290	—	—	—	5,090	37,400	—	—	—	35,800
1.900	4.50	—	—	H-40	.225	1.613	1.519	—	—	—	—	—	7,770	7,630	—	—	—	—	52,000	—	—	—	—
2.063	—	—	3.25	J-55	.156	1.751	1.657	—	—	—	1.657	2.325	7,690	7,280	—	—	—	7,000	51,400	—	—	—	49,300
2.063	4.50	—	—	L-80	.225	1.613	1.519	—	—	—	—	—	10,690	10,500	—	—	—	—	71,400	—	—	—	—
2.063	—	—	3.25	L-80	.156	1.751	1.657	—	—	—	1.657	2.325	11,180	10,590	—	—	—	10,180	74,800	—	—	—	71,700
2.063	4.50	—	—	N-80	.225	1.613	1.519	—	—	—	—	—	15,550	15,270	—	—	—	—	103,900	—	—	—	—
2.063	—	—	3.25	C-90	.156	1.751	1.657	—	—	—	1.657	2.325	11,180	10,590	—	—	—	10,180	74,800	—	—	—	71,700
2.063	4.50	—	—	C-90	.225	1.613	1.519	—	—	—	—	—	15,550	15,270	—	—	—	—	103,900	—	—	—	—
2.063	—	—	3.25	T-95	.156	1.751	1.657	—	—	—	1.657	2.325	12,420	11,910	—	—	—	11,460	84,200	—	—	—	80,700
2.063	4.50	—	—	T-95	.225	1.613	1.519	—	—	—	—	—	17,490	17,180	—	—	—	—	116,900	—	—	—	—
2.063	4.50	—	—	T-95	.225	1.751	1.657	—	—	—	1.657	2.325	12,980	12,570	—	—	—	12,090	88,800	—	—	—	85,100
2.063	4.50	—	—	P-110	.225	1.613	1.519	—	—	—	—	—	18,460	18,130	—	—	—	—	123,400	—	—	—	—
2.063	4.50	—	—	P-110	.225	1.613	1.519	—	—	—	—	—	21,380	20,990	—	—	—	—	142,900	—	—	—	—
2.375	4.00	—	—	H-40	.167	2.041	1.947	2.875	—	—	—	—	5,230	4,920	4,920	—	—	—	46,300	30,100	—	—	—
2.375	4.60	4.70	—	H-40	.190	1.995	1.901	2.875	3.063	2.910	—	—	5,890	5,600	5,600	5,600	5,600	—	52,200	35,900	52,200	52,200	—
2.375	4.00	—	—	J-55	.167	2.041	1.947	2.875	—	—	—	—	7,190	6,770	6,770	—	—	—	63,700	41,400	—	—	—
2.375	4.60	—	—	J-55	.190	1.995	1.901	2.875	3.063	2.910	—	—	8,100	7,700	7,700	7,700	7,700	—	71,700	49,400	71,700	71,700	—
2.375	4.00	—	—	L-80	.167	2.041	1.947	2.875	3.063	2.910	—	—	9,980	9,840	9,840	9,840	—	—	92,600	60,200	—	—	—
2.375	4.60	4.70	—	L-80	.190	1.995	1.901	2.875	3.063	2.910	—	—	11,780	11,200	11,200	11,200	11,200	—	104,300	71,800	104,300	104,300	—
2.375	5.80	5.95	—	L-80	.254	1.867	1.773	2.875	3.063	2.910	—	—	15,280	14,970	14,970	14,860	11,440	—	135,400	102,900	135,400	135,400	—
2.375	6.60	—	—	L-80	.295	1.785	1.691	—	—	—	—	—	17,410	17,390	—	—	—	—	154,200	—	—	—	—
2.375	7.35	7.45	—	L-80	.336	1.703	1.609	—	3.063	—	—	—	19,430	19,810	—	14,860	11,440	—	172,200	60,200	172,200	141,300	—
2.375	4.00	—	—	N-80	.167	2.041	1.947	2.875	3.063	2.910	—	—	9,980	9,840	9,840	—	—	—	92,600	60,200	—	—	—
2.375	4.60	4.70	—	N-80	.190	1.995	1.901	2.875	3.063	2.910	—	—	11,780	11,200	11,200	11,200	11,200	—	104,300	71,800	104,300	104,300	—
2.375	5.80	5.95	—	N-80	.254	1.867	1.773	2.875	3.063	2.910	—	—	15,280	14,970	14,970	14,860	11,440	—	135,400	102,900	135,400	135,400	—
2.375	4.00	—	—	C-90	.167	2.041	1.947	2.875	3.063	2.910	—	—	10,940	11,070	11,070	—	—	—	104,200	67,700	104,300	104,300	—
2.375	4.60	4.70	—	C-90	.190	1.995	1.901	2.875	3.063	2.910	—	—	13,250	12,600	12,600	12,600	12,600	—	117,400	80,800	117,400	117,400	—
2.375	5.80	5.95	—	C-90	.254	1.867	1.773	2.875	3.063	2.910	—	—	17,190	16,840	16,840	16,720	12,870	—	152,300	115,700	152,300	152,300	—
2.375	6.60	—	—	C-90	.295	1.785	1.691	—	—	—	—	—	19,580	19,560	—	—	—	—	173,500	—	—	—	—
2.375	7.35	7.45	—	C-90	.336	1.703	1.609	—	3.063	—	—	—	21,860	22,280	—	16,720	12,870	—	193,700	—	193,700	159,000	—
2.375	4.00	—	—	T-95	.167	2.041	1.947	2.875	3.063	2.910	—	—	11,410	11,690	—	13,300	13,300	—	110,000	85,300	123,900	123,900	—
2.375	4.60	4.70	—	T-95	.190	1.995	1.901	2.875	3.063	2.910	—	—	13,980	13,300	13,300	13,300	13,300	—	123,900	85,300	123,900	123,900	—
2.375	5.80	5.95	—	P-110	.254	1.867	1.773	2.875	3.063	2.910	—	—	18,150	17,780	17,780	17,650	13,580	—	160,700	122,200	160,700	160,700	—
2.375	6.60	—	—	P-110	.295	1.785	1.691	—	—	—	—	—	20,670	20,650	—	—	—	—	183,200	—	—	—	—
2.375	7.35	7.45	—	J-55	.336	1.703	1.609	—	3.063	—	—	—	23,080	23,520	—	17,650	13,580	—	204,400	—	204,400	167,800	—
2.375	4.60	4.70	—	C-90	.190	1.995	1.901	2.875	3.063	2.910	—	—	16,130	15,400	15,400	15,400	15,400	—	143,400	98,800	143,400	143,400	—
2.375	5.80	5.95	—	T-95	.254	1.867	1.773	2.875	3.063	2.910	—	—	21,010	20,590	20,590	20,430	15,730	—	186,100	141,500	186,100	186,100	—
2.875	6.40	6.50	—	T-95	.217	2.441	2.347	3.500	3.668	3.460	—	—	5,580	5,280	5,280	5,280	5,280	—	72,500	52,700	72,500	72,500	—
2.875	6.40	6.50	—	P-110	.217	2.441	2.347	3.500	3.668	3.460	—	—	7,680	7,260	7,260	7,260	7,260	—	99,700	72,500	99,700	99,700	—
2.875	6.40	6.50	—	H-40	.217	2.441	2.347	3.500	3.668	3.460	—	—	11,170	10,570	10,570	10,570	10,570	—	145,000	105,400	145,000	145,000	—
2.875	7.80	7.90	—	L-80	.276	2.323	2.229	3.500	3.668	3.460	—	—	13,890	13,440	13,440	13,440	11,030	—	180,300	140,700	180,300	180,300	—
2.875	8.60	8.70	—	L-80	.308	2.259	2.165	3.500	3.668	3.460	—	—	15,300	15,000	15,000	14,940	11,030	—	198,700	159,200	198,700	193,100	—

Table 3.12 shows minimum tubing performance properties.

TABLE 3.12—MINIMUM PERFORMANCE PROPERTIES OF TUBING (continued)

Size OD, D (in.)	Nominal Weight Threaded and Coupled Non-upset (lbm/ft)	Threaded and Coupled Upset (lbm/ft)	Integral Joint (lbm/ft)	Grade	Wall Thickness, t (in.)	d (in.)	Drift Diameter (in.)	Coupling OD Non-upset W (in.)	Upset Regular W (in.)	Upset Special Clearance Wc (in.)	Integral Joint Drift Diameter (in.)	Integral Joint OD of Box (in.)	Collapse Resistance	Plain End	Internal Yield Pressure Nonupset (psi)	Threaded and Coupled Upset Regular (psi)	Upset Special Clearance (psi)	Integral Joint (psi)	Pipe Body Yield (lbf)	Joint Yield Strength Nonupset (lbf)	Threaded and Coupled Upset Regular (lbf)	Upset Special Clearance (lbf)	Integral Joint (lbf)
1	2	3	4	5	6	7	8	9	10	11	12	13	14	15	16	17	18	19	20	21	22	23	24
2.875	9.35	9.45	—	L-80	.340	2.195	2.101	—	3.668	3.460	—	—	16,680	16,560	—	14,940	11,030	—	216,600	—	216,600	193,100	—
2.875	10.50	—	—	L-80	.392	2.091	1.997	—	—	—	—	—	18,840	19,090	—	—	—	—	244,600	—	—	—	—
2.875	11.50	—	—	L-80	.440	1.995	1.901	—	—	—	—	—	20,740	21,430	—	—	—	—	269,300	—	—	—	—
2.875	6.40	6.50	—	N-80	.217	2.441	2.347	3.500	3.668	3.460	—	—	11,170	10,570	10,570	10,570	10,570	—	145,000	105,400	145,000	145,000	—
2.875	7.80	7.90	—	N-80	.276	2.323	2.229	3.500	3.668	3.460	—	—	13,890	13,440	13,440	13,440	11,030	—	180,300	140,700	180,300	180,300	—
2.875	8.60	8.70	—	N-80	.308	2.259	2.165	3.500	3.668	3.460	—	—	15,300	15,000	15,000	14,940	11,030	—	198,700	159,200	198,700	193,100	—
2.875	6.40	6.50	—	C-90	.217	2.441	2.347	3.500	3.668	3.460	—	—	12,380	11,890	11,890	11,890	11,890	—	163,100	118,600	163,100	163,100	—
2.875	7.80	7.90	—	C-90	.276	2.323	2.229	3.500	3.668	3.460	—	—	15,620	15,120	15,120	15,120	12,410	—	202,900	158,300	202,900	202,900	—
2.875	8.60	8.70	—	C-90	.308	2.259	2.165	3.500	3.668	3.460	—	—	17,220	16,870	16,870	16,810	12,410	—	223,600	179,100	223,600	217,300	—
2.875	9.35	9.45	—	C-90	.340	2.195	2.101	—	3.668	3.460	—	—	18,770	18,630	—	16,810	12,410	—	243,700	—	243,700	217,300	—
2.875	10.50	—	—	C-90	.392	2.091	1.997	—	—	—	—	—	21,200	21,470	—	—	—	—	275,200	—	—	—	—
2.875	11.50	—	—	C-90	.440	1.995	1.901	—	—	—	—	—	23,330	24,100	—	—	—	—	302,900	—	—	—	—
2.875	6.40	6.50	—	T-95	.217	2.441	2.347	3.500	3.668	3.460	—	—	12,940	12,550	12,550	12,550	12,550	—	172,100	125,200	172,100	172,100	—
2.875	7.80	7.90	—	T-95	.276	2.323	2.229	3.500	3.668	3.460	—	—	16,490	15,960	15,960	15,960	13,100	—	214,100	167,100	214,100	214,100	—
2.875	8.60	8.70	—	T-95	.308	2.259	2.165	3.500	3.668	3.460	—	—	18,170	17,810	17,810	17,740	13,100	—	236,000	189,100	236,000	229,400	—
2.875	9.35	9.45	—	T-95	.340	2.195	2.101	—	3.668	3.460	—	—	19,810	19,660	—	17,740	13,100	—	257,300	—	257,300	229,440	—
2.875	10.50	—	—	T-95	.392	2.091	1.997	—	—	—	—	—	22,370	22,670	—	—	—	—	290,500	—	—	—	—
2.875	11.50	—	—	T-95	.440	1.995	1.901	—	—	—	—	—	24,630	25,440	—	—	—	—	319,800	—	—	—	—
2.875	6.40	6.50	—	P-110	.217	2.441	2.347	3.500	3.668	3.460	—	—	14,550	14,530	14,530	14,530	14,530	—	199,300	145,000	199,300	199,300	—
2.875	7.80	7.90	—	P-110	.276	2.323	2.229	3.500	3.668	3.460	—	—	19,090	18,480	18,480	18,480	15,160	—	247,900	193,500	247,900	247,900	—
2.875	8.60	8.50	—	P-110	.308	2.259	2.165	3.500	3.668	3.460	—	—	21,040	20,620	20,620	20,540	15,160	—	273,200	218,900	273,200	265,600	—
3.500	7.70	—	—	H-40	.216	3.068	2.943	4.250	—	—	—	—	4,630	4,320	4,320	—	—	—	89,100	65,000	—	—	—
3.500	9.20	9.30	—	H-40	.254	2.992	2.867	4.250	4.500	4.180	—	—	5,380	5,080	5,080	5,080	5,080	—	103,600	79,400	103,600	103,600	—
3.500	10.20	—	—	H-40	.289	2.922	2.797	4.250	—	—	—	—	6,060	5,780	5,780	—	—	—	116,600	92,500	—	—	—
3.500	7.70	—	—	J-55	.216	3.068	2.943	4.250	—	—	—	—	5,970	5,940	5,940	—	—	—	122,500	89,400	—	—	—
3.500	9.20	9.30	—	J-55	.254	2.992	2.867	4.250	4.500	4.180	—	—	7,400	6,990	6,990	6,990	6,990	—	142,500	109,200	142,500	142,500	—
3.500	10.20	—	—	J-55	.289	2.922	2.797	4.250	—	—	—	—	8,330	7,950	7,950	—	—	—	160,300	127,200	—	—	—
3.500	7.70	—	—	L-80	.216	3.068	2.943	4.250	—	—	—	—	7,870	8,640	8,640	—	—	—	178,200	130,000	—	—	—
3.500	9.20	9.30	—	L-80	.254	2.992	2.867	4.250	4.500	4.180	—	—	10,540	10,160	10,160	10,160	10,160	—	207,200	158,900	207,200	207,200	—
3.500	10.20	—	—	L-80	.289	2.922	2.797	4.250	—	—	—	—	12,120	11,560	11,560	—	—	—	233,200	185,000	—	—	—
3.500	12.70	12.95	—	L-80	.375	2.750	2.625	4.250	4.500	4.180	—	—	15,310	15,000	15,000	15,000	10,660	—	294,600	246,200	294,600	273,100	—
3.500	14.30	—	—	L-80	.430	2.640	2.515	—	—	—	—	—	17,240	17,200	—	—	—	—	331,800	—	—	—	—
3.500	15.50	—	—	L-80	.476	2.548	2.423	—	—	—	—	—	18,800	19,040	—	—	—	—	361,800	—	—	—	—
3.500	17.00	—	—	L-80	.530	2.440	2.315	—	—	—	—	—	20,560	21,200	—	—	—	—	395,600	—	—	—	—
3.500	7.70	—	—	N-80	.216	3.068	2.943	4.250	—	—	—	—	7,870	8,640	8,640	—	—	—	178,200	130,000	—	—	—
3.500	9.20	9.30	—	N-80	.254	2.992	2.867	4.250	4.500	4.180	—	—	10,540	10,160	10,160	10,160	10,160	—	207,200	158,900	207,200	207,200	—
3.500	10.20	—	—	N-80	.289	2.922	2.797	4.250	—	—	—	—	12,120	11,560	11,560	—	—	—	233,200	185,000	—	—	—
3.500	12.70	12.95	—	N-80	.375	2.750	2.625	4.250	4.500	4.180	—	—	15,310	15,000	15,000	15,000	10,660	—	294,600	246,200	294,600	273,100	—
3.500	7.70	—	—	C-90	.216	3.068	2.943	4.250	—	—	—	—	8,540	9,720	9,720	—	—	—	200,500	146,300	—	—	—
3.500	9.20	9.30	—	C-90	.254	2.992	2.867	4.250	4.500	4.180	—	—	11,570	11,430	11,430	11,430	11,430	—	233,100	178,700	233,100	233,100	—
3.500	10.20	—	—	C-90	.289	2.922	2.797	4.250	—	—	—	—	13,640	13,010	13,010	—	—	—	262,400	208,100	—	—	—
3.500	12.70	12.95	—	C-90	.375	2.750	2.625	4.250	4.500	4.180	—	—	17,220	16,880	16,880	16,880	11,990	—	331,400	277,000	331,400	307,300	—
3.500	14.30	—	—	C-90	.430	2.640	2.515	—	—	—	—	—	19,400	19,350	—	—	—	—	373,200	—	—	—	—
3.500	15.50	—	—	C-90	.476	2.548	2.423	—	—	—	—	—	21,150	21,420	—	—	—	—	407,000	—	—	—	—
3.500	17.00	—	—	C-90	.530	2.440	2.315	—	—	—	—	—	23,130	23,850	—	—	—	—	445,100	—	—	—	—
3.500	7.70	—	—	T-95	.216	3.068	2.943	4.250	—	—	—	—	8,850	10,260	10,260	—	—	—	211,700	154,400	—	—	—
3.500	9.20	9.30	—	T-95	.254	2.992	2.867	4.250	4.500	4.180	—	—	12,080	12,070	12,070	12,070	12,070	—	246,000	188,700	246,000	246,000	—
3.500	10.20	—	—	T-95	.289	2.922	2.797	4.250	—	—	—	—	14,390	13,730	13,730	—	—	—	276,900	219,600	—	—	—
3.500	12.70	12.95	—	T-95	.375	2.750	2.625	4.250	4.500	4.180	—	—	18,180	17,810	17,810	17,810	12,660	—	349,800	292,400	349,800	324,300	—
3.500	14.30	—	—	T-95	.430	2.640	2.515	—	—	—	—	—	20,480	20,430	—	—	—	—	394,000	—	—	—	—
3.500	15.50	—	—	T-95	.476	2.548	2.423	—	—	—	—	—	22,330	22,610	—	—	—	—	429,600	—	—	—	—
3.500	17.00	—	—	T-95	.530	2.440	2.315	—	—	—	—	—	24,410	25,170	—	—	—	—	469,800	—	—	—	—

Table 3.12 shows minimum tubing performance properties.

TABLE 3.12—MINIMUM PERFORMANCE PROPERTIES OF TUBING (continued)

1	2	3	4	5	6	7	8	9	10	11	12	13	14	15	16	17	18	19	20	21	22	23	24
	Nominal Weight							Threaded and Coupled							Internal Yield Pressure					Joint Yield Strength (F_j)			
	Threaded and Coupled							Coupling Outside Diameter			Integral Joint				Threaded and Coupled					Threaded and Coupled			
									Upset							Upset					Upset		
Size OD, D (in.)	Non-upset (lbm/ft)	Upset (lbm/ft)	Integral Joint (lbm/ft)	Grade	Wall Thickness, t (in.)	d (in.)	Drift Diameter (in.)	Non-upset W (in.)	Regular W (in.)	Special Clearance W_c (in.)	Drift Diameter (in.)	OD of Box (in.)	Collapse Resistance	Plain End	Nonupset (psi)	Regular (psi)	Special Clearance (psi)	Integral Joint (psi)	Pipe Body Yield (lbf)	Nonupset (lbf)	Regular (lbf)	Special Clearance (lbf)	Integral Joint (lbf)
3.500	9.20	9.30	—	P-110	.254	2.992	2.867	4.250	4.500	4.180	—	—	13,530	13,970	13,970	13,970	13,970	—	284,900	218,500	284,900	284,900	—
3.500	12.70	12.95	—	P-110	.375	2.750	2.625	4.250	4.500	4.180	—	—	21,050	20,630	20,630	20,630	14,660	—	405,000	338,600	405,000	375,500	—
4.000	9.50	—	—	H-40	.226	3.548	3.423	4.750	—	—	—	—	4,050	3,960	3,960	—	—	—	107,200	72,000	—	—	—
4.000	—	11.00	—	H-40	.262	3.476	3.351	—	5.000	—	—	—	4,900	4,590	—	4,590	—	—	123,100	—	123,100	—	—
4.000	9.50	—	—	J-55	.226	3.548	3.423	4.750	—	—	—	—	5,110	5,440	5,440	—	—	—	147,400	99,000	—	—	—
4.000	—	11.00	—	J-55	.262	3.476	3.351	—	5.000	—	—	—	6,590	6,300	—	6,300	—	—	169,200	—	169,200	—	—
4.000	9.50	—	—	L-80	.226	3.548	3.423	4.750	—	—	—	—	6,590	7,910	7,910	—	—	—	214,400	144,000	—	—	—
4.000	—	11.00	—	L-80	.262	3.476	3.351	—	5.000	—	—	—	8,800	9,170	—	9,170	—	—	246,200	—	246,200	—	—
4.000	13.20	—	—	L-80	.330	3.340	3.215	—	—	—	—	—	12,110	11,550	—	—	—	—	304,400	—	—	—	—
4.000	16.10	—	—	L-80	.415	3.170	3.045	—	—	—	—	—	14,880	14,530	—	—	—	—	373,900	—	—	—	—
4.000	18.90	—	—	L-80	.500	3.000	2.875	—	—	—	—	—	17,500	17,500	—	—	—	—	439,800	—	—	—	—
4.000	22.20	—	—	L-80	.610	2.780	2.655	—	—	—	—	—	20,680	21,350	—	—	—	—	519,800	—	—	—	—
4.000	9.50	—	—	N-80	.226	3.548	3.423	4.750	—	—	—	—	6,590	7,910	7,910	—	—	—	214,400	144,000	—	—	—
4.000	—	11.00	—	N-80	.262	3.476	3.351	—	5.000	—	—	—	8,800	9,170	—	9,170	—	—	246,200	—	246,200	—	—
4.000	9.50	—	—	C-90	.226	3.548	3.423	4.750	—	—	—	—	7,080	8,900	8,900	—	—	—	241,200	162,000	—	—	—
4.000	—	11.00	—	C-90	.262	3.476	3.351	—	5.000	—	—	—	9,590	10,320	—	10,320	—	—	276,900	—	276,900	—	—
4.000	13.20	—	—	C-90	.330	3.340	3.215	—	—	—	—	—	13,620	12,990	—	—	—	—	342,500	—	—	—	—
4.000	16.10	—	—	C-90	.415	3.170	3.045	—	—	—	—	—	16,740	16,340	—	—	—	—	420,700	—	—	—	—
4.000	18.90	—	—	C-90	.500	3.000	2.875	—	—	—	—	—	19,690	19,690	—	—	—	—	494,800	—	—	—	—
4.000	22.20	—	—	C-90	.610	2.780	2.655	—	—	—	—	—	23,260	24,020	—	—	—	—	584,700	—	—	—	—
4.000	9.50	—	—	T-95	.226	3.548	3.423	4.750	—	—	—	—	7,310	9,390	9,390	—	—	—	254,600	171,000	—	—	—
4.000	—	11.00	—	T-95	.262	3.476	3.351	—	5.000	—	—	—	9,980	10,890	—	10,890	—	—	292,300	—	292,300	—	—
4.000	13.20	—	—	T-95	.330	3.340	3.215	—	—	—	—	—	14,380	13,720	—	—	—	—	361,500	—	—	—	—
4.000	16.10	—	—	T-95	.415	3.170	3.045	—	—	—	—	—	17,670	17,250	—	—	—	—	444,000	—	—	—	—
4.000	18.90	—	—	T-95	.500	3.000	2.875	—	—	—	—	—	20,780	20,780	—	—	—	—	522,300	—	—	—	—
4.000	22.20	—	—	T-95	.610	2.780	2.655	—	—	—	—	—	24,560	25,350	—	—	—	—	617,200	—	—	—	—
4.500	12.60	12.75	—	H-40	.271	3.958	3.833	5.200	5.563	—	—	—	4,490	4,220	4,220	4,220	—	—	144,000	104,400	144,000	—	—
4.500	12.60	12.75	—	J-55	.271	3.958	3.833	5.200	5.563	—	—	—	5,730	5,800	5,800	5,800	—	—	198,000	143,500	198,000	—	—
4.500	12.60	12.75	—	L-80	.271	3.958	3.833	5.200	5.563	—	—	—	7,500	8,430	8,430	8,430	—	—	288,000	208,700	288,000	—	—
4.500	15.20	—	—	L-80	.337	3.826	3.701	—	—	—	—	—	11,080	10,480	—	—	—	—	352,600	—	—	—	—
4.500	17.00	—	—	L-80	.380	3.740	3.615	—	—	—	—	—	12,370	11,820	—	—	—	—	393,400	—	—	—	—
4.500	18.90	—	—	L-80	.430	3.640	3.515	—	—	—	—	—	13,830	13,380	—	—	—	—	439,800	—	—	—	—
4.500	21.50	—	—	L-80	.500	3.500	3.375	—	—	—	—	—	15,800	15,560	—	—	—	—	502,600	—	—	—	—
4.500	23.70	—	—	L-80	.560	3.380	3.255	—	—	—	—	—	17,430	17,420	—	—	—	—	554,600	—	—	—	—
4.500	26.00	—	—	L-80	.630	3.240	3.115	—	—	—	—	—	19,260	19,600	—	—	—	—	612,800	—	—	—	—
4.500	12.60	12.75	—	N-80	.271	3.958	3.833	5.200	5.563	—	—	—	7,500	8,430	8,430	8,430	—	—	288,000	208,700	288,000	—	—
4.500	12.60	12.75	—	C-90	.271	3.958	3.833	5.200	5.563	—	—	—	8,120	9,490	9,490	9,490	—	—	324,000	234,800	324,000	—	—
4.500	15.20	—	—	C-90	.337	3.826	3.701	—	—	—	—	—	12,220	11,800	—	—	—	—	396,600	—	—	—	—
4.500	17.00	—	—	C-90	.380	3.740	3.615	—	—	—	—	—	13,920	13,300	—	—	—	—	442,600	—	—	—	—
4.500	18.90	—	—	C-90	.430	3.640	3.515	—	—	—	—	—	15,560	15,050	—	—	—	—	494,800	—	—	—	—
4.500	21.50	—	—	C-90	.500	3.500	3.375	—	—	—	—	—	17,780	17,500	—	—	—	—	565,500	—	—	—	—
4.500	23.70	—	—	C-90	.560	3.380	3.255	—	—	—	—	—	19,610	19,600	—	—	—	—	623,900	—	—	—	—
4.500	26.00	—	—	C-90	.630	3.240	3.115	—	—	—	—	—	21,670	22,050	—	—	—	—	689,400	—	—	—	—
4.500	12.60	12.75	—	T-95	.271	3.958	3.833	5.200	5.563	—	—	—	8,410	10,010	10,010	10,010	—	—	342,000	247,900	342,000	—	—
4.500	15.20	—	—	T-95	.337	3.826	3.701	—	—	—	—	—	12,760	12,450	—	—	—	—	418,700	—	—	—	—
4.500	17.00	—	—	T-95	.380	3.740	3.615	—	—	—	—	—	14,690	14,040	—	—	—	—	467,200	—	—	—	—
4.500	18.90	—	—	T-95	.430	3.640	3.515	—	—	—	—	—	16,420	15,890	—	—	—	—	522,300	—	—	—	—
4.500	21.50	—	—	T-95	.500	3.500	3.375	—	—	—	—	—	18,770	18,470	—	—	—	—	596,900	—	—	—	—
4.500	23.70	—	—	T-95	.565	3.380	3.255	—	—	—	—	—	20,700	20,690	—	—	—	—	658,500	—	—	—	—
4.500	26.00	—	—	T-95	.630	3.240	3.115	—	—	—	—	—	22,880	23,280	—	—	—	—	727,700	—	—	—	—

Table 3.12 shows minimum tubing performance properties.

4.4 Perforating

Flow Efficiency of Perforated Systems

Well Flow Efficiency

$$J = \frac{q_p}{q_{oh}} = \frac{\ln(r_e / r_w)}{\left[\ln(r_e / r_w) + s_t\right]}$$

J Productivity ratio
q_p Flow from perforated system, bbl/D
q_{oh} Flow from ideal open hole, bbl/D
r_e Drainage radius, ft
r_w Wellbore radius, ft
s_t Total skin

Pressure Drop Across Perforations

$$\Delta P_{\text{perf}} = \frac{0.237 \rho Q^2}{C^2 N^2 H^4}$$

ΔP_{perf} Perforation pressure drop, psi
ρ Fluid density, lbf/gal
Q Injection rate, bbl/min
C Discharge coefficient
N Total number of perforation holes
H Perforation-hole diameter, in.

4.5 Acidizing

Fracture Gradient

$$FG = \frac{ISIP + 0.052 \rho_f D}{D}$$

FG Fracture gradient, psi/ft
$ISIP$ Instantaneous shut-in pressure obtained from a fracturing job, psi
ρ_f Fluid density at the time of the ISIP, lbf/gal
D Datum depth (normally midperforation), ft

Maximum Injection Rate

$$q_{i,\,\text{max}} = \frac{4.917 \times 10^{-6} kh\left(p_{bd} - \bar{p}\right)}{\mu \ln\left(\dfrac{0.472 r_e}{r_w} + s\right)}$$

$q_{i,\,\text{max}}$ Maximum injection rate, STB/D
p_{bd} Breakdown pressure, psi
$\bar{p}$ Average reservoir pressure, psi
k Permeability, md
h Reservoir thickness, ft
μ Viscosity, cp
r_e Drainage radius, ft
r_w Well radius, cp
s Skin effects, dimensionless

Maximum Tubing-Injection Pressure

$$p_{ti,\,max} = p_{bd} - \Delta p_{PE} + \Delta p_F$$

$$p_{bd} = FG \times D$$

$$\Delta p_{PE} = 0.433 \times \gamma_F \times D$$

$$\Delta p_F = \frac{7.51 q_i^2 D \gamma_F}{d^5 N_{Re}^{\;0.25}}$$

$$N_{Re} = 132,714.3 \frac{q_i \gamma_F}{d\mu}$$

q_i	Injection rate, bbl/min
p_{bd}	Breakdown pressure, psi
$p_{ti,\,max}$	Maximum tubing-injection pressure, psi
Δp_{PE}	Potential energy pressure drop, psi
Δp_F	Frictional pressure drop, psi
N_{Re}	Reynolds Number
FG	Fracture gradient, psi/ft
γ_F	Fluid specific gravity
μ	Viscosity, cp
D	Depth, ft
d	Tubular ID, in.

4.6 Fracturing

Basic Hydraulic Fracturing Equations

Breakdown Pressure

$$p_b = 3\sigma_h - \sigma_H + T - p_o$$

Reopening Pressure

$$p_r = -\sigma_H + 3\sigma_h$$

Propagation Pressure

$$p_p = \sigma_h + \Delta p_k + \Delta p_p$$

Instantaneous Shut-in Pressure

$$ISIP = \sigma_h + \delta_p$$

Closure Pressure

$$p_c = \sigma_h$$

σ_H	Major geostatic horizontal stress
σ_h	Geostatic horizontal stress
T	Minimum tensile strength, psi
P_o	Pore pressure, psi
Δp_k	Cohesive resistance of the material
Δp_p	Pressure drop in the fracture
δ_p	Extra pressure needed to keep the fracture open after shut in

General Fracturing Treatment Formulas

Bottomhole Fracturing Pressure Gradient

$$G_F = \frac{ISIP + P_h}{D}$$

Bottomhole Fracturing Pressure

$$P_F = G_F \times D$$

P_F Bottomhole fracturing pressure, psi
G_F Bottomhole fracturing pressure gradient, psi/ft

Instantaneous Shut-Down Pressure

$$P_i = P_F - P_h$$

P_i Instantaneous shut-down pressure, psi

Total Surface Pressure

$$P_S = P_F + P_{tf} + P_{pf} - P_h$$

P_S Total surface pressure, psi

Hydraulic Horsepower

$$HHP = 0.0245 \times P_S \times Q$$

P_h Total hydrostatic pressure, psi
P_{pf} Perforation friction pressure, psi
P_{tf} Total tubular friction pressure, psi
Q Injection rate, bbl/min
D Depth of producing interval, ft

Net Pressure

$$p_n = p_f - \sigma_{min}$$

p_n Net pressure, m/Lt^2
p_f Actual pressure in the fracture, m/Lt^2
σ_{min} Minimum horizontal stress, m/Lt^2

Dimensionless Fracture Conductivity

$$C_{fD} = \frac{k_f w}{k x_f}$$

C_{fD} Dimensionless fracture conductivity
k_f Fracture permeability, md
k Permeability, md
w Width, ft
x_f Productive fracture half-length, ft

Optimum Fracture Conductivity

$$C_{fD} = 31.4159 k L_f$$

C_{fD} Dimensionless fracture conductivity
k Formation permeability, md
L_f Fracture half-length, ft

Fracture Productivity Index, Prats' Method

$$\frac{J}{J_o} = \frac{\ln\left(\frac{r_e}{r_w}\right)}{\ln\left(\frac{r_e}{0.5L_f}\right)}$$

J Productivity index, STB/D/psi
J_o Productivity index of unfractured well, STB/D/psi
r_e Drainage radius, ft
r_w Wellbore radius, ft
L_f Fracture half-length, ft

Perforation Friction

$$\Delta p_{pf} = 0.2369 \frac{q^2 \rho}{n^2 D_p^4 C^2}$$

Δp_{pf} Pressure drop caused by perforation friction, psi
q Total flow rate, BPM
ρ Fluid density, lbm/ft³
n Number of perforations
D_p Perforation diameter, in.
C Discharge coefficient, usually 0.9

Mass Dissolving Power of an Acid

$$X_C = \frac{\rho_c \beta C}{\rho_{CaCO_3}}$$

$$\beta = \frac{(\text{molecular weight of rock})(\text{rock stoichiometric coefficient})}{(\text{molecular weight of acid})(\text{acid stoichiometric coefficient})}$$

X_C Dissolving power of acid
β Dissolving power coefficient
ρ_C Density of acid solution, lbm/ft³
ρ_{CaCO_3} Density of calcium carbonate, lbm/ft³
C Weight fraction concentration

Hydraulic Power

$$H_h = 0.0245 p_s q_t$$

H_h Hydraulic power, hp
p_s Surface injection pressure, psi
q_t Flow rate, bbl/min

Surface Injection Pressure

$$p_s = p_t + \Delta p_f + \Delta p_p + \Delta p_s$$

p_t Bottomhole fracture treating pressure, $= G_f D$
Δp_f Frictional pressure drop in the pipe, psi
Δp_p Pressure drop in the perforation, psi
Δp_s Hydrostatic pressure, psi
D Depth, ft
G_f Fracture gradient, psi/ft

Hydrostatic Pressure

$$\rho = 8.34\gamma + \frac{\chi}{1+0.0456\chi}$$

$$\gamma_T = \gamma_{60}\left[1 - \beta(T-60)\right]$$

$$\Delta p_s = 0.052\rho_T D$$

ρ	Density of the fluid/sand mix, ppg
γ	Specific gravity of fluid
χ	Concentration of sand, ppg
β	Thermal coefficient of expansion fluid
T	Temperature, °F

4.7 Well Production Problems

Low Productivity

Determine Permeability Based on Flow Regime

Radial Flow

$$\bar{k} = -\frac{qB\mu}{4\pi H_R m_R}$$

Linear Flow

$$k_y = \frac{\mu}{\pi\phi c_t}\left(\frac{qB}{m_L H_L X_L}\right)^2$$

k	Average permeability, md
q	Volumetric liquid production rate, bbl/D
B	Formation volume factor
μ	Fluid viscosity
c_t	Total reservoir compressibility
ϕ	Initial reservoir pressure
m_R	Slope of plot of bottomhole pressure vs. time on a semilog scale
m_L	Slope of plot of bottomhole pressure vs. square root of time
H_R	Thickness of the radial flow (h or L)
H_L	For linear flow $= h$, for pseudolinear flow $= h - z_w$
X_L	For linear flow $= 2x_f$, for pseudolinear flow $= L$
z_w	Distance between the horizontal well and the z boundary
L	Length of the horizontal well

Skin Effect

$$S = S_D + S_{C+\theta} + S_P + \sum S_{PS}$$

S	General expression of the skin factor
S_D	Damage skin during cementing, well completion, fluid injection, and oil and gas production
$S_{C+\theta}$	Skin component resulting from partial completion and deviation angle
S_P	Skin component resulting from the nonideal flow condition around the perforations associated with casedhole completion
$\sum S_{PS}$	Pseudoskin components resulting from non-Darcy flow effect, multiphase effect, and flow convergence near the wellbore

Coiled Tubing

Hydrostatic Pressure Test

$$P = \frac{2 \times f \times Y \times t_{min}}{D}$$

Internal Yield Pressure

$$P = \frac{2 \times Y \times t_{min}}{D}$$

Torsional Yield Strength

$$T_f = \frac{Y \times \left[D^4 - \left(D - 2t_{min} \right)^4 \right]}{7.113 \times D}$$

T_f	Torsional yield strength, in./lbf
P	Hydrostatic test pressure, psi
f	Test factor = 0.80
Y	Specified minimum yield strength, psi
t_{min}	Minimum specified wall thickness of the thinnest wall segment of tubing on the spool, mm or in. $\approx 0.95t$
D	Specified OD, in.

Pressure Losses

Inside Casing and Tubing

$$\Delta P_i = \frac{L\rho^{0.8} V^{1.8} \mu^{0.2}}{3,212,923 D^{1.2}} = \frac{L\rho^{0.8} Q^{1.8} \mu^{0.2}}{10,141 D^{4.8}}$$

In Annulus

$$\Delta P_a = \frac{L\rho^{0.8} V^{1.8} \mu^{0.2}}{2,519,939 \left(D_o - D_i \right)^{1.2}} = \frac{L\rho^{0.8} Q^{1.8} \mu^{0.2}}{7,952 \left(D_o + D_i \right)^{1.8} \left(D_o - D_i \right)^3}$$

ΔP_i	Internal pressure loss, psi
L	Length, ft
V	Circulation velocity, ft/min
Q	Flow rate, gal/min
D	Inner diameter, in.
D_o, D_i	Outer and inner diameter, respectively, in.
μ	Dynamic viscosity, cp
ρ	Density of fluid, lbm/gal

Water Coning

Schols Critical Production Rate

$$q_c = \left[\frac{\left(\rho_w - \rho_o \right) k_o \left(h^2 - D^2 \right)}{2,049 \mu_o B_o} \right] \left[0.432 + \frac{\pi}{\ln \left(r_e / r_w \right)} \right] \left(\frac{h}{r_e} \right)^{0.14}$$

q_c	Critical production rate, STB/D
ρ_o	Density of oil, g/cc
ρ_w	Density of water, g/cc
h	Thickness, ft
D	Thickness of perforated interval, ft
k_o	Effective oil permeability, md

μ_o Oil viscosity, cp
B_o Oil formation volume factor, bbl/STB
r_e Drainage radius, ft
r_w Wellbore radius, ft

Gilbert's Choke Equation

$$P_{tf} = \left(435 R_{gL}^{0.546} q_t\right) / S^{1.89}$$

P_{tf} Flowing tubing pressure, psig
R_{gL} Gas/liquid ratio, 10^3 scf/bbl
q_t Gross liquid rate (oil and water), bbl/D
S Choke size in 1/64 in.

4.8 Artificial Lift

Total Dynamic Head

$$TDH = WFL + F_t + F_{fl} + E_v + P_{\text{vessel}}$$

TDH Total dynamic head, ft
WFL Working fluid level, ft
F_t Friction loss in tubing, ft
F_{fl} Friction loss in flowline, ft
E_v Elevation change, ft
P_{vessel} Vessel pressure, ft

Sucker Rod Pumping

$$PRL_{max} = W_f + (0.9 + F_1) W_r$$

$$PRL_{min} = (0.9 - F_2) W_r$$

$$W_f = S_f (62.4) \frac{D_r A_p}{144}$$

$$W_r = \frac{\gamma_s D_r A_r}{144}$$

$$W_{rf} = W_r L (1 - 0.128 G)$$

$$F_0 = 0.34 \times G \times D^2 \times H$$

$$CBE = 1.06(W_{rf} + \tfrac{1}{2} F_o)$$

$$F_1 = \frac{SN^2 \left(1 \pm \dfrac{c}{h}\right)}{70,471.2} \qquad \text{For conventional and air-balanced units}$$

$$F_2 = \frac{SN^2 \left(1 \mp \dfrac{c}{h}\right)}{70,471.2} \qquad \text{For conventional and air-balanced units}$$

$$C = 0.5 W_f + W_r \left(0.9 \pm \frac{SN^2}{70,471.2} \frac{c}{h}\right) \qquad \text{For conventional and air-balanced units}$$

$$T = \frac{1}{4} S \left(W_f + \frac{2 SN^2 W_r}{70,471.2}\right)$$

$$N_{\text{limit}} = \frac{187.7}{\sqrt{S\left(1 \mp \dfrac{c}{h}\right)}} \qquad \text{For conventional and air-balanced units}$$

$$q = 0.1484\left(A_p N S_p E_v\right)$$

$$\delta l_r = \frac{W_f D_r}{A_r E}$$

$$S_p = S - \frac{12D}{E}\left[W_f\left(\frac{1}{A_r} + \frac{1}{A_t}\right) - \frac{SN^2\left(1 \pm \dfrac{c}{h}\right)}{70,471.2}\frac{W_r}{A_r}\right]$$

$$q_s = \frac{k_p}{\mu}\frac{\left(d_b - d_p\right)^{2.9}\left(d_b + d_p\right)}{d_b^{0.1}}\frac{\Delta p}{L_p}$$

$$P_{pm} = F_s\left(P_h + P_f\right)$$

$$P_f = 6.31 \times 10^{-7} W_r S N$$

$$P_h = 7.36 \times 10^{-6} q \gamma_l L_N$$

$$L_N = H + \frac{p_{tf}}{0.433\gamma_l}$$

$PRL_{\max}$	Maximum polished rod load, lbm
$PRL_{\min}$	Minimum polished rod load, lbm
W_f	Fluid load, lbm
W_r	Weight of the rod string in air, lbm/ft
W_{rf}	Weight of the rod string in fluid, lbm
L	Depth of rod string
G	Specific gravity of fluid
F_o	Gross plunger load, lbm
CBE	Counter-weight required, lbm
S_f	Specific gravity of fluid in tubing
S	Polished rod stroke length, in.
k_p	Constant
T	Peak torque, lbm
q	Pump deliverability (liquid production rate), bbl/D
q_s	Slippage rate, bbl/D
D	Plunger diameter, in.
D_r	Length of rod string, ft
d_b	Plunger OD, in.
d_p	Barrel ID, in.
L_p	Length of plunger, in.
L_N	Net lift, ft
Δp	Barrel ID, in.
P_h	Hydraulic power, hp
P_{pm}	Prime mover power, hp
P_f	Power to overcome friction losses, hp
p_{tf}	Flowing tubinghead pressure, psi
F_s	Safety factor = 1.25 to 1.5
N_{limit}	Speed limit of pumping stroke
C	Counter-balance load, lbm

γ_s Specific weight of steel, (490 lbm/ft^3)
γ_l Liquid specific gravity, water = 1
A_p Gross plunger cross-sectional area, in.^{2}
A_r Sucker rod cross-sectional area, in.^{2}
A_t Tubing cross-sectional area, in.^{2}
N Number of pumping strokes per minute
c Length of crank arm, ft
μ Viscosity, cp
H Depth to the average fluid level in the annulus, ft
h Length of pitman arm, ft
S_p Effective plunger stroke length, in.
E_v Volumetric efficiency of the plunger
E Modulus of elasticity of steel, (30×10^6 lbf/in.^{2})

Beam-Pump HP

Hydraulic HP (H_{HP})

$$H_{HP} = \left[Q(\text{BFPD}) \times 350(\text{lbf/bbl}) \times G \times H(\text{ft}) \right] / \left\{ 33{,}000 \left[(\text{lbf-ft})/(\text{HP-min}) \right] \times 1{,}440(\text{min/D}) \right\}$$

or

$$H_{HP} = \left[F_o(\text{lbf}) \times S_p(\text{in.}) \times N(\text{spm}) \right] / \left\{ 12(\text{in./ft}) \times 33{,}000 \left[(\text{lbf-ft})/(\text{HP-min}) \right] \right\}$$

Polished-Rod HP $\left(P_{HP} \right)$

$$P_{HP} = \left[\text{card area (in.}^2) \times \text{dynamometer constant} \times S\ (\text{in.}) \times N\ (\text{spm}) \right] / (\text{card length} \times 12 \times 33{,}000)$$

Gear-Reducer HP $\left(G_{HP} \right)$

$$G_{HP} = (\text{gearbox-torque rating}) / 4{,}960$$

V-Belt-Drive HP $\left(V_{HP} \right)$

$$V_{HP} = [\text{peak crankshaft torque (lbf-in.)} \times N(\text{spm})] / 70{,}000$$

Brake HP $\left(B_{HP} \right)$

$$B_{HP} = P_{HP} / (\text{efficiency factor})$$

Additionally, a minimum estimate for B_{HP} by National Electrical Manufacturers Association (NEMA) for Design D and C motors is as follows:

$$B_{HP} = [Q(\text{BFPD}) \times L(\text{ft})] / (\text{derating factor})$$

Indicated HP $\left(I_{HP} \right)$

$$I_{HP} = B_{HP} / (\text{derating factor})$$

Q Fluid flow rate, B/D
G Specific gravity of the combined fluid in the tubing
H Pump seating depth, ft
F_o Differential fluid load on full pump-plunger cross-sectional area, lbf
S Surface stroke length, in.
S_p Downhole pump-plunger stroke length, in.
N Pumping-unit speed, spm
L Pump-seating nipple depth, ft

Rod Number	Plunger Diameter (in.)	Rod Weight, W_r (lbm/ft)	Elastic Constant, E_r	Frequency Factor, F_e	Rod String (% of each size)				
					1	$\frac{7}{8}$	$\frac{3}{4}$	$\frac{5}{8}$	$\frac{1}{2}$
44	All	0.726	1.990×10^{-6}	1.000	–	–	–	–	100.0
54	1.06	0.908	1.668×10^{-6}	1.138	–	–	–	44.6	55.4
54	1.25	0.929	1.633×10^{-6}	1.140	–	–	–	49.5	50.5
54	1.50	0.957	1.584×10^{-6}	1.137	–	–	–	56.4	43.6
54	1.75	0.990	1.525×10^{-6}	1.122	–	–	–	64.6	35.4
54	2.00	1.027	1.460×10^{-6}	1.095	–	–	–	73.7	26.3
54	2.25	1.067	1.391×10^{-6}	1.061	–	–	–	83.4	16.6
54	2.5	1.108	1.318×10^{-6}	1.023	–	–	–	93.5	6.5
55	All	1.135	1.270×10^{-6}	1.000	–	–	–	100.0	–
64	1.06	1.164	1.382×10^{-6}	1.229	–	–	33.3	33.1	33.5
64	1.25	1.211	1.319×10^{-6}	1.215	–	–	37.2	35.9	26.9
64	1.50	1.275	1.232×10^{-6}	1.184	–	–	42.3	40.4	17.3
64	1.75	1.341	1.141×10^{-6}	1.145	–	–	47.4	45.2	7.4
65	1.06	1.307	1.138×10^{-6}	1.098	–	–	34.4	65.6	–
65	1.25	1.321	1.127×10^{-6}	1.104	–	–	37.3	62.7	–
65	1.50	1.343	1.110×10^{-6}	1.110	–	–	41.8	58.2	–
65	1.75	1.369	1.090×10^{-6}	1.114	–	–	46.9	53.1	–

Table 4.1a—Pump unit design (Bradley 1987).

Rod Number	Plunger Diameter (in.)	Rod Weight, W_r (lbm/ft)	Elastic Constant, E_r	Frequency Factor, F_e	Rod String (% of each size)				
					1	$\frac{7}{8}$	$\frac{3}{4}$	$\frac{5}{8}$	$\frac{1}{2}$
65	2.00	1.394	1.070×10^{-6}	1.114	–	–	52.0	48.0	–
65	2.25	1.426	1.045×10^{-6}	1.110	–	–	58.4	41.6	–
65	2.50	1.460	1.018×10^{-6}	1.099	–	–	65.2	34.8	–
65	2.75	1.497	0.990×10^{-6}	1.082	–	–	72.5	27.5	–
65	3.25	1.574	0.930×10^{-6}	1.037	–	–	88.1	11.9	–
66	All	1.634	0.883×10^{-6}	1.000	–	–	100.0	–	–
75	1.06	1.566	0.997×10^{-6}	1.191	–	27.0	27.4	45.6	–
75	1.25	1.604	0.978×10^{-6}	1.193	–	29.4	29.8	40.8	–
75	1.50	1.664	0.935×10^{-6}	1.189	–	33.3	33.3	33.3	–
75	1.75	1.732	0.892×10^{-6}	1.174	–	37.8	37.0	25.1	–
75	2.00	1.803	0.847×10^{-6}	1.151	–	42.4	41.3	16.3	–
75	2.25	1.875	0.801×10^{-6}	1.121	–	46.9	45.8	7.2	–
76	1.06	1.802	0.816×10^{-6}	1.072	–	28.5	71.5	–	–
76	1.25	1.814	0.812×10^{-6}	1.077	–	30.6	69.4	–	–
76	1.50	1.833	0.804×10^{-6}	1.082	–	33.8	66.2	–	–
76	1.75	1.855	0.795×10^{-6}	1.088	–	37.5	62.5	–	–
76	2.00	1.880	0.785×10^{-6}	1.093	–	41.7	58.3	–	–
76	2.25	1.908	0.774×10^{-6}	1.096	–	46.5	53.5	–	–
76	2.50	1.934	0.764×10^{-6}	1.097	–	50.8	49.2	–	–
76	2.75	1.967	0.751×10^{-6}	1.094	–	56.5	43.5	–	–
76	3.25	2.039	0.722×10^{-6}	1.078	–	68.7	31.3	–	–
76	3.75	2.119	0.690×10^{-6}	1.047	–	82.3	17.7	–	–

Table 4.1a cont'd—Pump unit design (Bradley 1987).

Rod Number	Plunger Diameter (in.)	Rod Weight, W_r (lbm/ft)	Elastic Constant, E_r	Frequency Factor, F_e	Rod String (% of each size)					
					1 ¼	1 ⅛	1	⅞	¾	⅝
77	All	2.224	0.649×10^{-6}	1.000	–	–	–	1.0	–	–
85	1.06	1.883	0.878×10^{-6}	1.261	–	–	22.2	22.4	22.4	33.0
85	1.25	1.943	0.841×10^{-6}	1.253	–	–	23.9	24.2	24.3	27.6
85	1.50	2.039	0.791×10^{-6}	1.232	–	–	26.7	27.4	26.8	19.2
85	1.75	2.138	0.738×10^{-6}	1.201	–	–	29.6	30.4	29.5	10.5
86	1.06	2.058	0.742×10^{-6}	1.151	–	–	22.6	23.0	54.3	–
86	1.25	2.087	0.732×10^{-6}	1.156	–	–	24.3	24.5	51.2	–
86	1.50	2.133	0.717×10^{-6}	1.162	–	–	26.8	27.0	46.3	–
86	1.75	2.185	0.699×10^{-6}	1.164	–	–	29.4	30.0	40.6	–
86	2.00	2.247	0.679×10^{-6}	1.161	–	–	32.8	33.2	33.9	–
86	2.25	2.315	0.656×10^{-6}	1.153	–	–	36.9	36.0	27.1	–
86	2.50	2.385	0.633×10^{-6}	1.138	–	–	40.6	39.7	19.7	–
86	2.75	2.455	0.610×10^{-6}	1.119	–	–	44.5	43.3	12.2	–
87	1.06	2.390	0.612×10^{-6}	1.055	–	–	24.3	75.7	–	–
87	1.25	2.399	0.610×10^{-6}	1.058	–	–	25.7	74.3	–	–
87	1.50	2.413	0.607×10^{-6}	1.062	–	–	27.7	72.3	–	–
87	1.75	2.430	0.603×10^{-6}	1.066	–	–	30.3	69.7	–	–
87	2.00	2.450	0.598×10^{-6}	1.071	–	–	33.2	66.8	–	–
87	2.25	2.472	0.594×10^{-6}	1.075	–	–	36.4	63.6	–	–
87	2.50	2.496	0.588×10^{-6}	1.079	–	–	39.9	60.1	–	–
87	2.75	2.523	0.582×10^{-6}	1.082	–	–	43.9	56.1	–	–
87	3.25	2.575	0.570×10^{-6}	1.084	–	–	51.6	48.4	–	–

Table 4.1b—Pump unit design, cont'd (Bradley 1987).

Rod Number	Plunger Diameter (in.)	Rod Weight, W_r (lbm/ft)	Elastic Constant, E_r	Frequency Factor, F_o	Rod String (% of each size)					
					1 ¼	1 ⅛	1	⅞	¾	⅝
87	3.75	2.641	0.556×10^{-6}	1.078	–	–	61.2	38.8	–	–
87	4.75	2.793	0.522×10^{-6}	1.038	–	–	83.6	16.4	–	–
88	All	2.904	0.497×10^{-6}	1.000	–	–	100.0	–	–	–
96	1.06	2.382	0.670×10^{-6}	1.222	–	19.1	19.2	19.5	42.3	–
96	1.25	2.485	0.655×10^{-6}	1.224	–	20.5	20.5	20.7	38.3	–
96	1.50	2.511	0.633×10^{-6}	1.223	–	22.4	22.5	22.8	32.3	–
96	1.75	2.607	0.606×10^{-6}	1.213	–	24.8	25.1	25.1	25.1	–
96	2.00	2.703	0.578×10^{-6}	1.196	–	27.1	27.9	27.4	17.6	–
96	2.25	2.806	0.549×10^{-6}	1.172	–	29.6	30.7	29.8	9.8	–
97	1.06	2.645	0.568×10^{-6}	1.120	–	19.6	20.0	60.3	–	–
97	1.25	2.670	0.563×10^{-6}	1.124	–	20.8	21.2	58.0	–	–
97	1.50	2.707	0.556×10^{-6}	1.131	–	22.5	23.0	54.5	–	–
97	1.75	2.751	0.548×10^{-6}	1.137	–	24.5	25.0	50.4	–	–
97	2.00	2.801	0.538×10^{-6}	1.141	–	26.8	27.4	45.7	–	–
97	2.25	2.856	0.528×10^{-6}	1.143	–	29.4	30.2	40.4	–	–
97	2.50	2.921	0.515×10^{-6}	1.141	–	32.5	33.1	34.4	–	–
97	2.75	2.989	0.503×10^{-6}	1.135	–	36.1	35.3	28.6	–	–
97	3.25	3.132	0.475×10^{-6}	1.111	–	42.9	41.9	15.2	–	–
98	1.06	3.068	0.475×10^{-6}	1.043	–	21.2	78.8	–	–	–
98	1.25	3.076	0.474×10^{-6}	1.045	–	22.2	77.8	–	–	–
98	1.50	3.089	0.472×10^{-6}	1.048	–	23.8	76.2	–	–	–
98	1.75	3.103	0.470×10^{-6}	1.051	–	25.7	74.3	–	–	–
98	2.00	3.118	0.468×10^{-6}	1.055	–	27.7	72.3	–	–	–
98	2.25	3.137	0.465×10^{-6}	1.058	–	30.1	69.9	–	–	–
98	2.5	3.157	0.463×10^{-6}	1.062	–	32.7	67.3	–	–	–
98	2.75	3.180	0.460×10^{-6}	1.066	–	35.6	64.4	–	–	–
98	3.25	3.231	0.453×10^{-6}	1.071	–	42.2	57.8	–	–	–
98	3.75	3.289	0.445×10^{-6}	1.074	–	49.7	50.3	–	–	–
98	4.75	3.412	0.428×10^{-6}	1.064	–	65.7	34.3	–	–	–
99	All	3.676	0.393×10^{-6}	1.000	–	100.0		–	–	–
107	1.06	2.977	0.524×10^{-6}	1.184	16.9	16.8	17.1	49.1	–	–
107	1.25	3.019	0.517×10^{-6}	1.189	17.9	17.8	18	46.8	–	–
107	1.50	3.085	0.506×10^{-6}	1.195	19.4	19.2	19.5	41.9	–	–
107	1.75	3.158	0.494×10^{-6}	1.197	21.0	21.0	21.2	36.9	–	–
107	2.00	3.238	0.480×10^{-6}	1.195	22.7	22.8	23.1	31.4	–	–
107	2.25	3.336	0.464×10^{-6}	1.187	25	25.0	25	25.0	–	–
107	2.50	3.435	0.447×10^{-6}	1.174	26.9	27.7	27.1	18.2	–	–
107	2.75	3.537	0.430×10^{-6}	1.156	29.1	30.2	29.3	11.3	–	–
108	1.06	3.325	0.447×10^{-6}	1.097	17.3	17.8	64.9	–	–	–
108	1.25	3.345	0.445×10^{-6}	1.101	18.1	18.6	68.2	–	–	–
108	1.50	3.376	0.441×10^{-6}	1.106	19.4	19.9	60.7	–	–	–
108	1.75	3.411	0.437×10^{-6}	1.111	20.9	21.4	57.7	–	–	–
108	2.00	3.452	0.432×10^{-6}	1.117	22.6	23.0	54.3	–	–	–
108	2.25	3.498	0.427×10^{-6}	1.121	24.5	25.0	50.5	–	–	–
108	2.50	3.548	0.421×10^{-6}	1.124	26.5	27.2	46.3	–	–	–
108	2.75	3.603	0.415×10^{-6}	1.126	28.7	29.6	41.6	–	–	–
108	3.25	3.731	0.400×10^{-6}	1.123	34.6	33.9	31.6	–	–	–
108	3.75	3.873	0.383×10^{-6}	1.108	40.6	39.5	19.9	–	–	–

Table 4.1b cont'd—Pump unit design (Bradley 1987).

Rod Number	Plunger Diameter (in.)	Rod Weight, W_r (lbm/ft)	Elastic Constant, E_r	Frequency Factor, F_e	Rod String (% of each size)					
					1 ¼	1 ⅛	1	⅞	¾	⅝
109	1.06	3.839	0.378×10^{-6}	1.035	18.9	81.1	–	–	–	–
109	1.25	3.845	0.378×10^{-6}	1.036	19.6	80.4	–	–	–	–
109	1.50	3.855	0.377×10^{-6}	1.038	20.7	79.3	–	–	–	–
109	1.75	3.867	0.376×10^{-6}	1.040	22.1	77.9	–	–	–	–
109	2.00	3.880	0.375×10^{-6}	1.043	23.7	76.3	–	–	–	–
109	2.25	3.896	0.374×10^{-6}	1.046	25.4	74.6	–	–	–	–
109	2.50	3.911	0.372×10^{-6}	1.048	27.2	72.8	–	–	–	–
109	2.75	3.93	0.371×10^{-6}	1.051	29.4	70.6	–	–	–	–
109	3.25	3.971	0.367×10^{-6}	1.057	34.2	65.8	–	–	–	–
109	3.75	4.020	0.363×10^{-6}	1.063	39.9	60.1	–	–	–	–
109	4.75	4.120	0.354×10^{-6}	1.066	51.5	48.5	–	–	–	–
1010	All	4.538	0.318×10^{-6}	1.000	100.0	–	–	–	–	–

Table 4.1b cont'd—Pump unit design (Bradley 1987).

Tubing Size	Nominal Weight T&C (lbm/ft)	Outside Diameter (in.)	Inside Diameter (in.)	Material Area in.^{2}	Elastic Constant E_t (in./lbm-ft)
1 ½	2.90	1.900	1.610	0.800	0.500×10^{-6}
2 ⅜	4.70	2.375	1.995	1.304	0.307×10^{-6}
2 ⅞	6.50	2.875	2.441	1.812	0.221×10^{-6}
3 ½	9.20	3.500	2.992	2.590	0.154×10^{-6}
4	11.00	4.000	3.476	3.077	0.130×10^{-6}
4 ½	12.75	4.500	3.958	3.601	0.111×10^{-6}

Table 4.2—Tubing data.

Rod Size	Metal Area, in.^{2}	Rod Weight in Air, W_r (lbm/ft)	Elastic Constant, E_r (in./lbm-ft)
½	0.196	0.72	1.99×10^{-6}
⅝	0.307	1.13	1.27×10^{-6}
¾	0.442	1.63	0.883×10^{-6}
⅞	0.601	2.22	0.649×10^{-6}
1	0.785	2.90	0.497×10^{-6}
1 ⅛	0.994	3.67	0.393×10^{-6}

Table 4.3—Sucker rod data.

Plunger Diameter, D (in.)	Plunger Diameter Squared, D^2 in.^{2}	Fluid Load Factor*, $(.340 \times D^2)$ (lbm/ft)	Pump Factor, $(.1166 \times D^2)$
1 1⁄16	1.1289	0.384	0.132
1 ¼	1.5625	0.531	0.182
1 ½	2.2500	0.765	0.262
1 ¾	3.0625	1.041	0.357
2	4.0000	1.360	0.466
2 ¼	5.0625	1.721	0.590
2 ½	6.2500	2.125	0.728
2 ¾	7.5625	2.571	0.881
3 ¾	14.0625	4.781	1.640
4 ¾	22.5625	7.671	2.63

*For fluids with specific gravity of 1.00

Table 4.4—Pump constants.

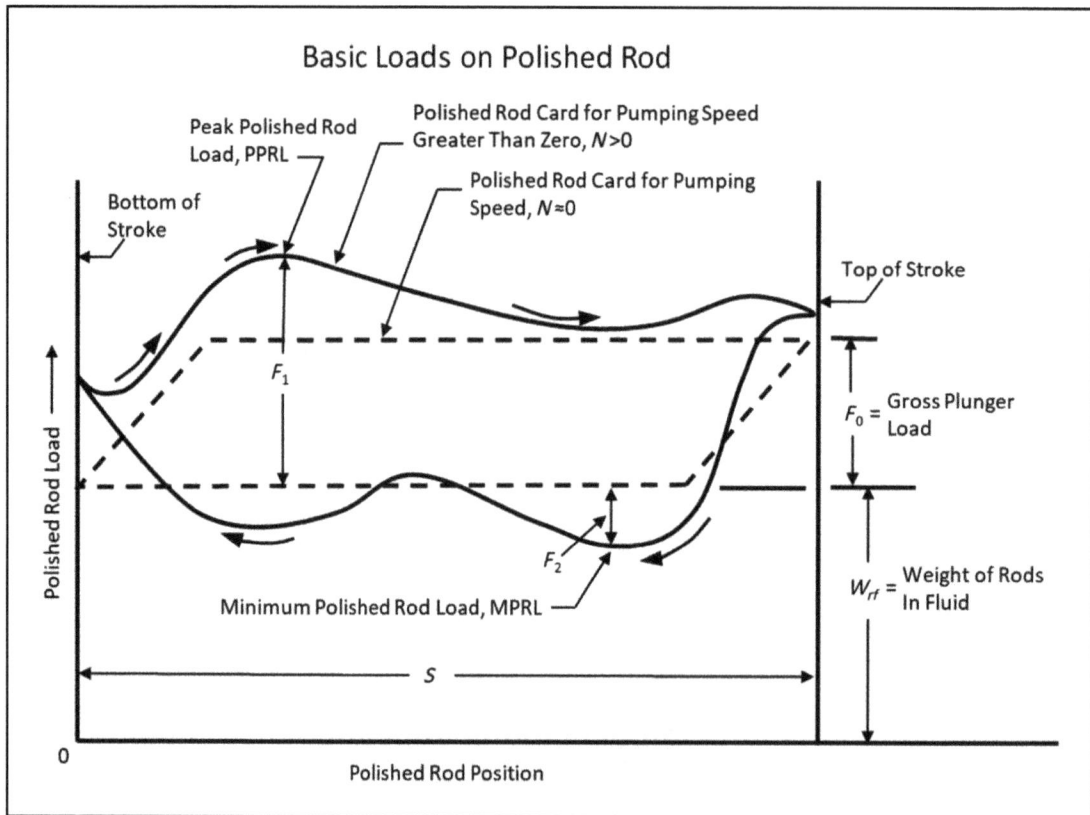

Fig. 4.1—Basic loads on polished rod (*API RP 11L* 1988).

Gas Lift

Depth of Top Gas-Lift Valve (D₁)

$$D_1 = \frac{\left(p_{ko} - p_{whu}\right)}{g_{sl}}$$

p_{ko} Kick-off injection gas pressure, psi
p_{whu} Wellhead U-tubing pressure, psi
g_{sl} Load fluid gradient, psi/ft

Gas-Lift Potential

$$GLR_{\text{opt},o} = GLR_{fm} + \frac{q_{g,\text{inj}}}{q_o}$$

$GLR_{\text{opt},o}$ Optimum GLR at operating flow rate, scf/STB
GLR_{fm} Formation oil GLR, scf/STB
$q_{g,\text{inj}}$ Lift gas injection rate, scf/D
q_o Expected operating liquid flow rate, scf/D

Electrical Submersible Pump (ESP)

$$h = \frac{\Delta p}{0.433}$$

h Pumping head, ft

Δp Pump pressure differential, psi

$$D_{\text{pump}} = D - \frac{p_{wf} - p_{\text{suction}}}{0.433\gamma_L}$$

D_{pump} Minimum pump depth, ft
D Depth of production interval, ft
p_{wf} Flowing bottomhole pressure, psia
p_{suction} Required suction pressure of pump, 150–300 psi
γ_L Specific gravity of production fluid, 1.0 for fresh water

Plunger Lift

$$GLR_{\min} = 400\frac{D}{1,000}$$

$GLR_{\min}$ Minimum required GLR by a rule of thumb, scf/bbl
D Depth of plunger, ft

$$p_c = p_{L\max} + \frac{p_{sh}}{f_{sl}}$$

p_c Required casing pressure, psia
$p_{L\max}$ Maximum line pressure, psia
p_{sh} Slug hydrostatic pressure, psia
f_{sl} Slug factor, 0.5 to 0.6

$$P_{c,\min} = \left[P_p + 14.7 + P_t + \left(P_{lh} + P_{lf}\right)V_{\text{slug}} \right]\left(1 + \frac{D}{K} \right)$$

$P_{c,\min}$ Required minimum casing pressure, psia
P_p $= W_p / A_t$, psia
P_t Tubinghead pressure, psia
P_{lh} Hydrostatic liquid gradient, psi/bbl slug
P_{lf} Fluid liquid gradient, psi/bbl slug
V_{slug} Slug volume, bbl
D Depth to plunger, ft
K Characteristic length for gas flow in tubing, ft

$$V_g = \frac{37.14 F_{gs} P_{c,\text{avg}} V_t}{z\left(T_{\text{avg}} + 460\right)}$$

V_g Required gas per cycle, Mscf
F_{gs} Modified Foss and Gaul slippage factor, $= 1 + 0.02 \times (D/1,000)$
$P_{c,\text{avg}}$ Average casing pressure, $= P_{c,\min}(1 + A_t/2A_a)$, psia
A_t Tubing cross-sectional area, in.^{2}
A_a Annulus cross-sectional area, in.^{2}
V_t Gas volume in tubing, $= A_t\left(D - V_{\text{slug}}L\right)$, Mscf
L Tubing inner capacity, ft/bbl
z Gas compressibility factor in average tubing condition
T_{avg} Average temperature in tubing, °F
V_{slug} Slug volume, bbl
D Depth to plunger, ft

$$N_{Cmax} = \frac{1,440}{\dfrac{D}{V_r} + \dfrac{D - V_{slug}L}{V_{fg}} + \dfrac{V_{slug}L}{V_{fl}}}$$

N_{Cmax}	Maximum number of cycles per day
V_{fg}	Plunger falling velocity in gas, ft/min
V_{fl}	Plunger falling velocity in liquid, ft/min
V_r	Plunger rising velocity, ft/min
V_{slug}	Slug volume, bbl
D	Depth to plunger, ft
L	Tubing inner capacity, ft/bbl

$$q_{Lmax} = N_{Cmax} V_{slug}$$

q_{Lmax}	Maximum liquid production rate, bbl/D
N_{Cmax}	Maximum number of cycles per day
V_{slug}	Slug volume, bbl

$$GLR_{min} = \frac{V_g}{V_{slug}}$$

GLR_{min}	Minimum required GLR, scf/bbl
V_g	Required gas per cycle, Mscf
V_{slug}	Slug volume, bbl

Hydraulic Jet Pumping

$$R = \frac{A_j}{A_t}$$

$$M = \frac{q_3}{q_1}$$

$$H = \frac{p_2 - p_3}{p_1 - p_2}$$

$$\eta = MH$$

$$q_1 = 1214.5 A_j \sqrt{\frac{p_1 - p_3}{\gamma_1}}$$

R	Dimensionless nozzle area
M	Dimensionless flow rate
H	Dimensionless head
η	Pump efficiency
p_1	Power fluid pressure, psia
q_1	Power fluid rate, bbl/D
p_2	Discharge pressure, psia
q_2	Total fluid rate in return column, $= q_1 + q_3$, bbl/D
p_3	Intake pressure, psia
q_3	Intake (produced) fluid rate, bbl/D
A_j	Jet nozzle area, in.^{2}
A_t	Total throat area, in.^{2}
γ_1	Specific gravity of the power fluid

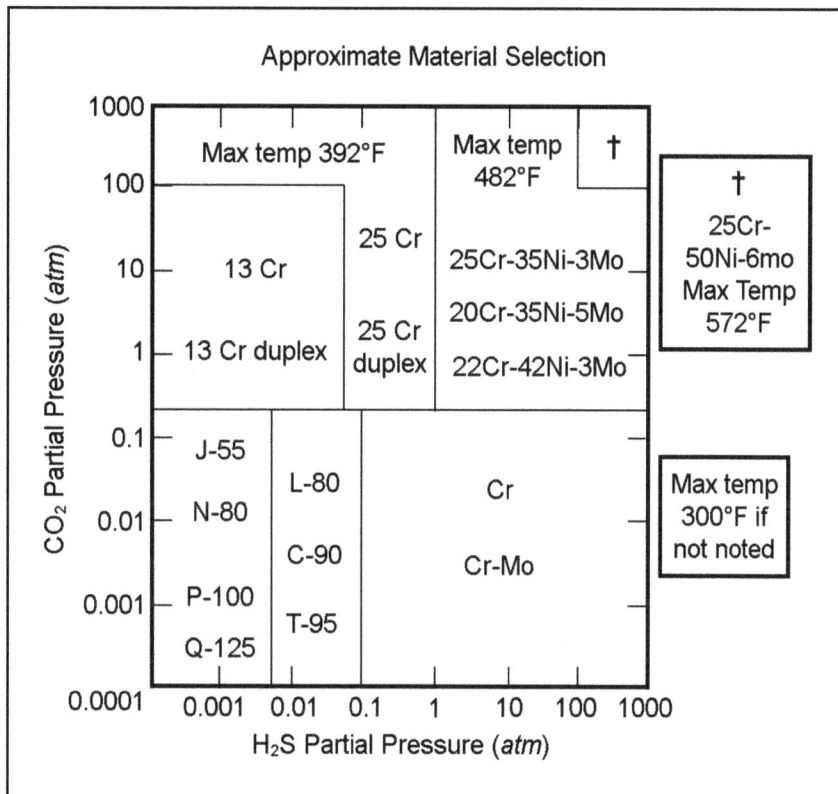

Fig. 4.2—Approximate material selection.

Scale

Calcium Carbonate Scale

Langelier Saturation Index (LSI) = $pH - pH_s$

$$pH_s = 0.1\log_{10}(TDS) - 13.12\log_{10}(T) - \log_{10}\left[(Hard)(Alk)\right] + 44.15$$

$$Hard = 1,000\left\{MW_{CaCO_s}\left([Ca]+[Mg]\right)\right\}$$

$$Alk = 500\left\{MW_{CaCO_s}\left(HCO_3\right) + 2[CO_3] + [OH] + [H]\right\}$$

$[H]$	is negligible
pH	Actual pH
pH_s	pH at saturation
TDS	Total dissolved solids, mg/L
T	Temperature, K
$Hard$	Hardness, mg/L
Alk	Alkalinity, mg/L
MW_{CaCO_s}	100.09 g/mol
$[ion]$	mol/L

Corrosion

Approximate Material Selectio

$$P_{CO_2} = z_{CO_2} P_{BH}$$

$$P_{H_2S} = P_{BH} \left(H_2S \text{ ppm} / 1 \times 10^6 \right)$$

Monitoring—Corrosion Coupon

$$CR = \frac{394 \times \Delta \text{mass}}{r_c A_s \Delta t}$$

$$\frac{\text{mil}}{\text{yr}} = \frac{\text{inch}}{1{,}000 \text{ yr}} = \frac{2.54 \text{ cm}}{1{,}000 \text{ yr}}$$

$$PR = d / \Delta t [=] \text{mil/yr}$$

P_{CO_2}	CO_2 partial pressure
z_{CO_2}	CO_2 fraction
P_{H_2S}	H_2S partial pressure
P_{BH}	Static bottomhole $P [=]$ atm
CR	Corrosion rate, $[=]$ thickness loss/time, mil/yr
Δmass	Mass lost, g
ρ_c	Coupon density, g/cm^3
A_s	Coupon surface area, cm^2
Δt	Test duration, yr
PR	Pitting rate, mil/yr
d	Deepest pit depth, mil

Tubing

OD (in.)	Nominal weight, T&C (upset) (lbm/ft)	Wall thickness (in.)	ID (in.)	Drift (in.)
1.9	2.9	0.145	1.610	1.516
2 ⅜	4.7	0.190	1.995	1.901
2 ⅜	5.95	0.254	1.867	1.773
2 ⅞	6.5	0.217	2.441	2.347
2 ⅞	7.9	0.276	2.323	2.229
2 ⅞	8.7	0.308	2.259	2.165
3 ½	9.3	0.254	2.992	2.867
3 ½	12.95	0.375	2.750	2.625
4	11	0.262	3.476	3.351
4 ½	12.75	0.271	3.958	3.833

Table 4.5—Tubing properties.

Nominal size (in.)	ID (in.)	Wall thickness (in.)	Approximate weight per foot (lbm/ft)	Capacity (bbl/ft)	Displacement (bbl/ft)
3.5	2.0625	0.719	25.3	0.0042	0.0092
3.5	2.25	0.625	23.2	0.005	0.0084
4	2.5625	0.719	27.2	0.0073	0.0100
4.2	2.75	0.875	41.0	0.0074	0.0149
5	3	1.000	49.3	0.0088	0.0179
5.5	3.375	1.063	57.0	0.0111	0.0207
6.625	4.5	1.063	70.8	0.0196	0.0257

Table 4.6—Tubing properties.

OD (in.)	Nominal Weight, T&C (Upset) (lbm/ft)	Grade	Collapse Resistance (psi)	Pipe Body Yield Strength, 1,000 (lbf)	Internal Pressure Resistance, Plain End (psi)
1.9	2.9	J-55	7,750	44	7,350
	2.9	L-80	11,280	64	10,680
	2.9	C-90	12,620	72	12,020
2 ⅜	4.7	J-55	8,100	72	7,700
	4.7	L-80	11,780	104	11,200
	5.95	L-80	15,280	135	14,970
	4.7	C-95	13,980	124	13,300
	5.95	C-95	18,150	161	17,780
	4.7	P-110	16,130	143	15,400
	5.95	P-110	21,010	186	20,590
2 ⅞	6.5	J-55	7,680	100	7,260
	6.5	L-80	11,170	145	10,570
	7.9	L-80	13,890	180	13,440
	8.7	L-80	15,300	199	15,000
	6.5	C-95	12,940	172	12,550
	7.9	C-95	16,490	214	15,960
	8.7	C-95	18,170	236	17,810
	6.5	P-110	14,550	199	14,530
	7.9	P-110	19,090	248	18,480
	8.7	P-110	21,040	273	20,620
3 ½	9.2	J-55	7,400	142	6,990
	9.2	L-80	10,540	207	10160
	12.95	L-80	15,310	295	15,000
	9.2	C-95	12,080	246	12,070
	12.95	C-95	18,180	350	17,810
	9.2	P-110	13,530	285	13,970
	12.95	P-110	21,050	405	20,630
4	11	J-55	6,590	169	6,300
	11	L-80	8,800	246	9,170
	11	C-95	9,980	292	10,890
	11	P-110	11,060	338	12,610
4 ½	12.75	J-55	5,730	198	5,800
	12.75	L-80	7,500	288	8,430
	12.75	C-95	8,410	342	10,010
	12.75	P-110	9,210	396	11,590

Table 4.7—Tubing properties.

Solution Densities

Specific Gravity	Percent CaCl₂ by Weight of		Weight of solution per		Pounds of CaCl₂ added to water per			Volume* of solution (bbl)
	Solution	Water	Gallon	Cubic foot	Gallon	Cubic foot	Barrel	
0.9982	0	0	8.33	62.32	–	–	–	1
1.0148	2	2.04	8.47	63.35	0.17	1.27	7.14	1.004
1.0316	4	4.17	8.61	64.40	0.347	2.60	14.59	1.008
1.0486	6	6.38	8.75	65.46	0.531	3.98	22.32	1.013
1.0659	8	8.70	8.89	66.54	0.725	5.42	30.44	1.019
1.0835	10	11.11	9.04	67.64	0.925	6.92	38.87	1.024
1.1015	12	13.64	9.19	68.71	1.136	8.50	47.72	1.030
1.1198	14	16.28	9.34	69.91	1.356	10.15	56.96	1.037

Table 4.8—Densities of CaCl₂ Solutions at 68°F.

Specific Gravity	Percent CaCl₂ by Weight of		Weight of solution per		Pounds of CaCl₂ added to water per			Volume* of solution (bbl)
	Solution	Water	Gallon	Cubic foot	Gallon	Cubic foot	Barrel	
1.1386	16	19.05	9.50	71.08	1.587	11.87	66.65	1.044
1.1578	18	21.95	9.66	72.28	1.828	13.68	76.79	1.052
1.1775	20	25.00	9.83	73.51	2.083	15.58	87.47	1.059
1.2284	25	33.33	10.25	76.69	2.776	20.77	116.61	1.084
1.2816	30	42.86	10.69	80.01	3.570	26.71	149.95	1.113
1.3373	35	53.85	11.16	83.48	4.486	33.56	188.40	1.148
1.3957	40	66.67	11.65	87.13	5.554	41.55	233.25	1.192

* Final volume of solution after adding specified quantity of calcium chloride to 1 bbl of fresh water.

Table 4.8 cont'd—Densities of CaCl₂ Solutions at 68°F.

Specific Gravity	Percent NaCl by Weight of		Weight of solution per		Pounds of NaCl added to water per			Volume* of solution (bbl)
	Solution	Water	Gallon	Cubic foot	Gallon	Cubic foot	Barrel	
0.9982	0	0	8.33	62.32	–	–	–	1
1.0053	1	1.01	8.39	62.76	0.084	0.63	3.53	1.003
1.0125	2	2.04	8.45	63.21	0.170	1.27	7.14	1.006
1.0268	4	4.17	8.57	64.10	0.347	2.60	14.59	1.013
1.0413	6	6.38	8.69	65.01	0.531	3.98	22.32	1.020
1.0559	8	8.70	8.81	65.92	0.725	5.42	30.44	1.028
1.0707	10	11.11	8.93	66.84	0.925	6.92	38.87	1.036
1.0857	12	13.64	9.06	67.78	1.136	8.50	47.72	1.045
1.1009	14	16.28	9.19	68.73	1.356	10.15	56.96	1.054
1.1162	16	19.05	9.31	69.68	1.587	11.87	66.65	1.065
1.1319	18	21.95	9.45	70.66	1.828	13.68	76.79	1.075
1.1478	20	25.00	9.58	71.65	2.083	15.58	87.47	1.087
1.1640	22	28.21	9.71	72.67	2.350	17.58	98.70	1.100
1.1804	24	31.58	9.85	73.69	2.631	19.68	110.49	1.113
1.1972	26	35.13	9.99	74.74	2.926	21.89	122.91	1.127

* Final volume of solution after adding specified quantity of sodium chloride to 1 bbl of fresh water.

Table 4.9—Densities of NaCl solutions at 68°F.

Salt, wt%	ppm	mg/L
0.5	5,000	5,020
1	10,000	10,050
2	20,000	20,250
3	30,000	30,700
4	40,000	41,100
5	50,000	52,000
6	60,000	62,500
7	70,000	73,000
8	80,000	84,500
9	90,000	95,000
10	100,000	107,100
11	110,000	118,500
12	120,000	130,300
13	130,000	142,000
14	140,000	154,100
15	150,000	166,500
16	160,000	178,600

Table 4.10—NaCl concentrations as wt%, ppm, and mg/L.

Salt, wt%	ppm	mg/L
17	170,000	191,000
18	180,000	203,700
19	190,000	216,500
20	200,000	229,600
21	210,000	243,000
22	220,000	256,100
23	230,000	270,000
24	240,000	283,300
25	250,000	297,200
26	260,000	311,300

Table 4.10 cont'd—NaCl concentrations as wt%, ppm, and mg/L.

Chapter 5

Facilities

5.1 Separation Units

Settling Theory

Horizontal Separators

Liquid Drops in Gas Phase

$$\frac{L_{\mathrm{eff}}d^2F_g}{h_g} = 421\frac{TZQ_g}{P}\left[\left(\left|\frac{\rho_g}{\rho_l - \rho_g}\right|\right)\frac{C_D}{d_m}\right]^{1/2}$$

d Vessel ID, in.
d_m Drop diameter, μm
h_g Gas-phase space height, in.
F_g Fractional gas cross-sectional area
L_{eff} Effective length of the vessel where separation occurs, ft
T Operating temperature, °R
Q_g Gas flow rate, MMscf/D
P Operating pressure, psia
Z Gas compressibility factor, dimensionless
ρ_l Liquid density, lbm/ft³
ρ_g Gas density, lbm/ft³
C_D Drag coefficient

Bubbles or Liquid Drops in Liquid Phase

$$\frac{L_{\mathrm{eff}}d^2F_c}{h_c} = \frac{Q_c}{12}\left[\left(\left|\frac{\rho_c}{\rho_d - \rho_c}\right|\right)\frac{C_D}{d_m}\right]^{1/2}$$

d_m Bubble or drop diameter, μm
h_c Continuous liquid-phase space height, in.
F_c Fractional continuous-phase cross-sectional area

ρ_d Dispersed liquid-phase density, lbm/ft³
ρ_c Continuous liquid-phase density, lbm/ft³
Q_c Continuous liquid-phase flow rate, B/D

$$h_c = \frac{0.00129 t_{rc}(\Delta\gamma)d_m^2}{\mu_c} \qquad \text{for low Reynolds number flow}$$

t_{rc} Continuous-phase retention time, min
μ_c Continuous-phase dynamic viscosity, cp
$\Delta\gamma$ Specific gravity difference (heavy/light) of continuous dispersed phases

Vertical Vessels

Liquid Drops in Gas Phase

$$d^2 = 5{,}054 \frac{TZQ_g}{P}\left[\left(\left|\frac{\rho_g}{\rho_l - \rho_g}\right|\right)\frac{C_D}{d_m}\right]^{1/2}$$

Bubbles or Liquid Drops in Liquid Phase

$$d^2 = Q_c\left[\left(\left|\frac{\rho_c}{\rho_d - \rho_c}\right|\right)\frac{C_D}{d_m}\right]^{1/2}$$

$$d^2 = 6{,}663\frac{Q_c\mu_c}{(\Delta\gamma)d_m^2} \qquad \text{for low Reynolds number flow}$$

d Vessel ID, in.
d_m Drop diameter, μm
T Operating temperature, °R
Q_g Gas flow rate, MMscf/D
P Operating pressure, psia
Z Gas compressibility factor, dimensionless
ρ_l Liquid density, lbm/ft³
ρ_g Gas density, lbm/ft³
C_D Drag coefficient
ρ_d Dispersed liquid-phase density, lbm/ft³
ρ_c Continuous liquid-phase density, lbm/ft³
Q_c Continuous liquid-phase flow rate, B/D
μ_c Continuous-phase dynamic viscosity, cp
$\Delta\gamma$ Specific gravity difference (heavy/light) of continuous and dispersed phases

Retention Time

Horizontal Vessels

$$d^2 L_{\text{eff}} = \frac{t_{ro}Q_o + t_{rw}Q_w}{1.4F_l}$$

Vertical Vessels

$$d^2(h_o + h_w) = \frac{t_{ro}Q_o + t_{rw}Q_w}{0.12}$$

d Vessel ID, in.
L_{eff} Effective length of the vessel where separation occurs, ft
t_{ro} Oil retention time, min
t_{rw} Water retention time, min
h_o Oil pad height, in.
h_w Water pad height, in.
Q_o Oil flow rate, B/D
Q_w Water flow rate, B/D
F_l Fraction of vessel cross-sectional area filled by liquid

Demister Sizing

$$V_m = K_d \sqrt{\frac{\rho_l - \rho_g}{\rho_g}}$$

Horizontal Vessels

$$A_d = \frac{0.327 \frac{TZQ_g}{P}}{K_d \sqrt{\frac{\rho_l - \rho_g}{\rho_g}}}$$

Vertical Vessels

$$d^2 = \frac{60 \frac{TZQ_g}{P}}{K_d \sqrt{\frac{\rho_l - \rho_g}{\rho_g}}}$$

d Vessel ID, in.
A_d Required demister area, in.^{2}
V_m Maximum velocity, ft/sec
K_d Demister capacity factor, ft/sec
ρ_l Liquid density, lbm/ft^3
ρ_g Gas density, lbm/ft^3
T Operating temperature, °R
Z Gas compressibility factor, dimensionless
Q_g Gas flow rate, MMscf/D
P Operating pressure, psia

Seam-to-Seam Length

Horizontal Vessels

$$L_{ss} = L_{eff} + \frac{d}{12} \text{(gas)} = \frac{4}{3} L_{eff} \text{ (liquid)}$$

Vertical Vessels

$$L_{ss} = \frac{h + \text{nozzle ID} + \text{demister height} + 54}{12} \quad \text{or}$$

$$L_{ss} = \frac{h + \text{nozzle ID} + d + \text{demister height} + 18}{12}$$

L_{ss} Seam-to-seam vessel length, ft
L_{eff} Effective length of the vessel where separation occurs, ft
h Height, ft
d Vessel ID, in.

Maximum Allowable Velocity (V_{max})

$$V_{max} = K_s \sqrt{(\rho_L - \rho_G)/\rho_G}$$

K_s Separator design parameter (see **Table 5.1**)
ρ_L Liquid-phase density, lbm/ft^3
ρ_G Gas-phase density, lbm/ft^3

Type	Height or Length, ft (m)	Typical K_s Range, ft/sec
Vertical	5 (1.52)	0.12 to 0.24
	10 (3.05)	0.18 to 0.35
Horizontal	10 (3.05	0.40 to 0.50
	Other lengths	0.40 to 0.50 $(L/10)^{0.56}$

Table 5.1—Maximum allowable velocity (*API SPEC 12J* 2008).

5.2 Treating and Processing Units

Emulsion Treating

$$\mu_e / \mu_o = 1 + 2.5f + 14.1f^2$$

μ_e Viscosity of emulsion, cp
μ_o Viscosity of clean oil, cp
f Fraction of the dispersed phase

$$Q = 16\Delta T \left(0.5 q_o \gamma_o + q_w \gamma_w \right)$$

Q Required heat input for an insulated vessel, Btu/hr
ΔT Temperature increase, °R
q_o Oil flow rate, B/D
γ_o Specific gravity of oil
q_w Water flow rate, B/D
γ_w Specific gravity of water

Gravity Settling

$$v = \frac{\left(1.78 \times 10^{-6}\right) \Delta \gamma_{ow} d^2}{\mu_o}$$

v Downward velocity of the water droplet relative to the oil, ft/s
d Water droplet diameter, μm
$\Delta \gamma_{ow}$ Specific gravity difference between the water and the oil (water/oil)
μ_o Dynamic viscosity of the oil, cp

Desalting

$$C_{so} = 0.35 C_{sw} \gamma_w f_w$$

C_{so} Salt content of the oil, lbm/1,000 bbl
C_{sw} Concentration of salt in produced water, ppm
γ_w Specific gravity of produced water
f_w Volume fraction of water in crude oil

Settling Tanks/Weir Height

$$h_{wd} = \left(h_{oo} - h_{ww}\right)\frac{\gamma_o}{\gamma_w} + h_{ww}$$

h_{wd} Height of water draw-off overflow nipple in the weir box above the tank bottom, ft
h_{oo} Height of clean oil outlet above the tank bottom, ft
h_{ww} Desired height of water wash in the tank above the tank bottom, ft
γ_o Specific gravity of oil
γ_w Specific gravity of water

Electrostatic Coalescing Treaters

$$F = \frac{70\varepsilon E^2 r^6}{d_i^4}$$

F Dipolar attractive force between drops, N
ε Dielectric constant, $C^2/N\bullet m^2$
E Electric field gradient, V/m
r Drop radius, m
d_i Interdrop distance, m

$$E_c \le \varepsilon \left(\frac{\sigma}{d}\right)^{0.5}$$

E_c Critical voltage gradient as instability limit, V/m
ε Dielectric constant, $C^2/N\bullet m^2$
σ Interfacial tension, N/m
d Droplet diameter, m

Inlet Spreader Design

$$q = AC_o\sqrt{2gh(\gamma_w - \gamma_o)}$$

Outlet Collectors Design

$$q = 19.65 C_o n D^2 \sqrt{h}$$

q Flow, ft^3/s
A Area of diffuser holes, ft^2
C_o Orifice factor, = 0.6 to 0.7
g Gravitational constant, = 32.2 ft/s^2
h Head, ft
γ_w Specific gravity of brine
γ_o Specific gravity of oil
D Hole diameter, in.
n Number of holes

Water Treating

Gravity Segregation

$$v = g_c \frac{\Delta\rho(d_p)^2}{18\mu_L}$$

v Velocity of the droplet or particle rising or settling in a continuous phase, cm/s
$\Delta\rho$ Difference in density of the dispersed particle and the continuous phase, g/cm^3
g_c Gravity acceleration constant, cm/s^2
d_p Dispersed particle diameter, cm
μ_L Viscosity of the continuous phase (liquid), $g/cm\bullet s$

Dispersion

$$d_{max} = 432 \left(\frac{t_r}{\Delta p}\right)^{2/5} \left(\frac{\sigma}{\rho_w}\right)^{3/5}$$

d_{max} Diameter of the droplet, above which only 5% of the oil volume is contained, μm
t_r Retention time, min
Δp Pressure drop, psi

σ Surface tension, dynes/cm

ρ_w Density, g/cm^3

Coalescence

$$t = \frac{(d_d)^4}{2 f_V K_s}$$

t Time for a small droplet to grow large in a quiet gravity-settling tank, seconds

d_d Oil droplet diameter, μm

f_V Volume fraction of the dispersed phase

K_s Empirical settling constant

Horizontal Pressure Vessel Sizing/Skim Vessels

$$d_i L_e = \frac{1{,}000 q_w \mu_w}{\Delta \gamma_{ow} (d_d)^2}$$

$$t_r = 0.7 \frac{(d_i)^2 L_e}{q_w}$$

Vertical Cylindrical Vessel/Skim Vessels

$$(d)^2 = 6{,}691 F \frac{q_w \mu_w}{\Delta \gamma_{ow} (d_d)^2}$$

d Vessel ID, in.

L_e Effective length in which separation occurs, ft, (for design use of 75% seam-to-seam length)

q_w Water flow rate, BWPD

μ_w Water viscosity, cp

$\Delta \gamma_{ow}$ Difference in specific gravity between oil and water

d_d Oil droplet diameter, μm

t_r Retention time, min

F A factor to account for turbulence and short circuiting

General Sizing Equation for Plate Coalescers

$$(d_d)^2 = \frac{4.7 q_w L_p \mu_w}{Z_p B_p L \Delta \gamma_{ow} \cos \theta}$$

$$(q_w)_{max} = \frac{1{,}562 Z_p B_p}{L_p}$$

d_d Design oil-droplet diameter, μm

q_w Bulk-water flow rate, BWPD

L_p Perpendicular distance between plates, in.

μ_w Water viscosity, cp

Z_p Height of the plate section perpendicular to the axis of water flow, ft

B_p Width of the plate section perpendicular to the axis of water flow, ft

L Length of the plate section parallel to the axis of water flow, ft

$\Delta \gamma_{ow}$ Difference in specific gravity between oil and water

θ Angle of the plate with the horizontal, degrees

Skim-Pile Sizing

$$(d_i)^2 L_{bs} = 19.1 \left(q_w + 0.356 A_d q_r + q_{WD} \right)$$

$$N_c = \frac{41.7 (d_i)^2 L_{bs}}{q_w t}$$

d_i Pile internal diameter, in.
L_{bs} Length of baffle section, ft
q_w Produced water rate if it is disposed in pile, B/D
A_d Deck area, ft^2
q_r Rainfall rate, in./hr
q_{WD} Washdown rate, bbl/D
N_c Number of nonflow cycles that a particle sees as it traverses the baffle section
t Time for the dump cycle, sec

Separating Suspended Solids From Produced Water by Gravity Settling

$$b = \frac{36 q_w \mu_w}{\Delta \gamma \left(d_p\right)^2 L_e}$$

$$v_w = 6.5 \times 10^{-5} \frac{q_w}{h_f b}$$

b Width (breath) of the flow channel, ft
q_w Water flow rate, B/D
μ_w Water viscosity, cp
γ Specific gravity
d_p Dispersed particle diameter, in.
L_e Effective length, ft
v_w Velocity of water, ft/s
h_f Height of the flume, ft

Separating Suspended Solids From Produced Water by Desanding Hydrocyclones

$$x_{98} = \frac{0.0094 D^{1.18} \exp(6.3c)}{Q_f^{0.45} \left(\rho_s - \rho_l\right)^{0.5}}$$

x_{98} Particle size at 98% efficiency, m
D ID of the cyclone, in.
c Solid concentration
Q_f Feed volumetric flow rate, m^3/sec
ρ_s Solid density, kg/m^3
ρ_l Liquid density, kg/m^3

Removing Dissolved Solids From Water by Membranes

$$Q_{pf} = K_f K_T K A \Delta p_{avg}$$

Q_{pf} Permeate flow, gal/D
K_f Fouling factor
K_T Membrane temperature-correction factor
K Permeate flow coefficient at standard temperature, gal/D-psi-ft^2
A   Membrane area, ft^2
Δp_{avg} Average transmembrane pressure drop, psi

Gas Treating

Estimating Solution-Circulation Rate

Circulation rate :   US gal/min $= 0.219 \times MW_a \times Q_g \times AG / \left(ML \times SG_a \times W_s\right)$

Acid-gas pick up :   scf/gal $= 31.72 \times ML \times SG_a \times W_s / MW_a$

MW_a Molecular weight of amine, lbm/mol
AG Percent acid gas, %

ML Mole loading, mol/mol

SG_a Specific gravity of amine

W_s Weight percent of solvent in solution, %

Q_g Gas flow rate, MMscf/D

Contactor-Design Consideration

$$V = K\left(\frac{\rho_l - \rho_g}{\rho_g}\right)^{0.5}$$

$$\rho_g = 2.70\frac{P\gamma}{ZT}$$

$$d = 7.75\left(\frac{Q_g TZ}{PV}\right)^{0.5}$$

V Superficial gas velocity, ft/sec

K Coefficient

ρ_l Liquid density, lbm/ft^3

ρ_g Gas density, lbm/ft^3

P Pressure, psia

T Absolute temperature, °R

Q_g Gas flow rate, MMscf/D

Z Gas compressibility factor, dimensionless

d ID, in.

γ Gas-specific gravity, air = 1

Gas Oil Ratio (GOR) at Primary Separator

GOR at primary separator, scf/separator bbl =

GOR for primary separator, scf/STB × oil shrinkage at primary separator, STB/separator bbl

5.3 Pumps

Hydrostatics

$$h = 2.31 p / \gamma$$

$$\gamma = \rho_f / \rho_w$$

h Height of the fluid column above a reference point

p Pressure, psia

γ Specific gravity of the liquid

ρ_f Density of the liquid being pumped, lbm/ft^3

ρ_w Density of water at standard conditions of temperature and pressure, lbm/ft^3

Hydrodynamics

Bernoulli's Equation

$$v_1^2 / 2g + p_1 / \rho + Z_1 = v_2^2 / 2g + p_2 / \rho + Z_2 + h_f$$

Velocity Head

$$v^2 / 2g = 0.00259 Q^2 / d^4$$

v Average velocity of the liquid in the pipe

g Acceleration of gravity

p Pressure

ρ Density, lbm/ft^3

Z Height above a datum, ft

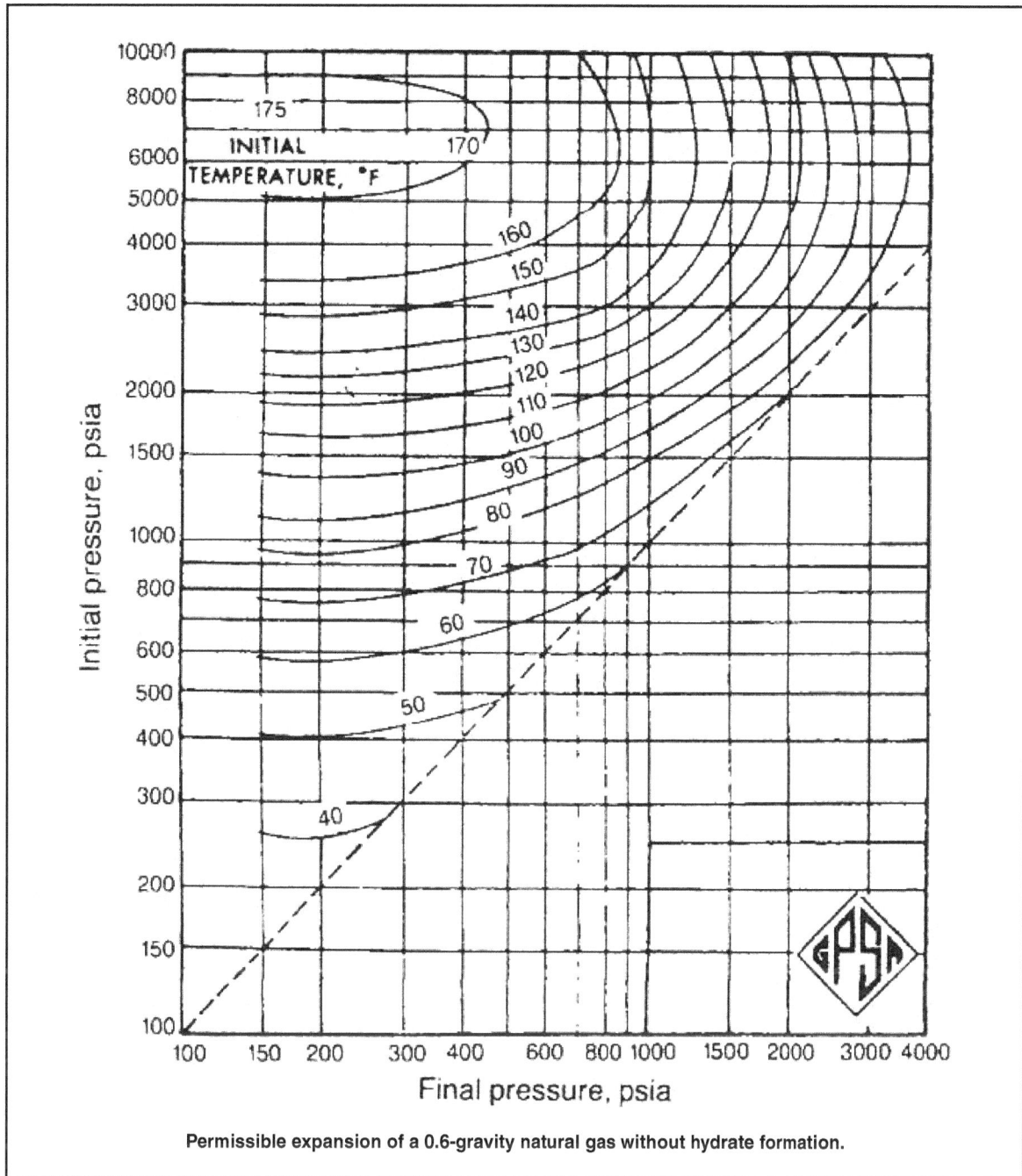

Permissible expansion of a 0.6-gravity natural gas without hydrate formation.

h_f Friction loss between points 1 and 2

d Pipe ID, in.

Q Flow rate, MMscf/D

Suction Head

$$H_s = \left[\left(p_1 - p_{f1} \right) \times 2.31 \right] / \gamma + H_1$$

H_s Suction head of liquid being pumped

p_1 Suction-vessel operating pressure

p_{f1} Pressure drop resulting from friction in the suction pipe

γ Specific gravity of liquid, dimensionless

H_1 Height of liquid suction vessel above pump reference point

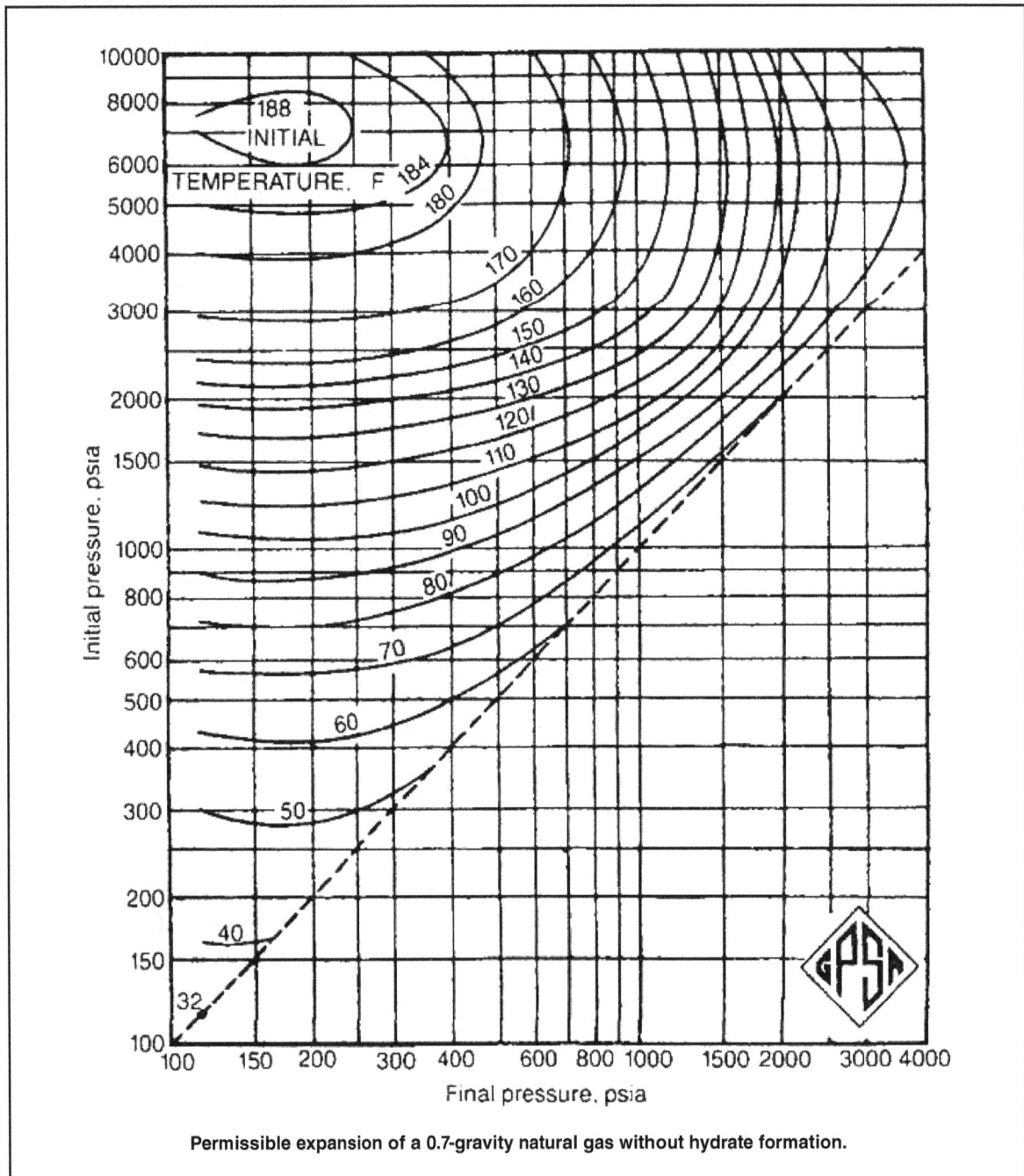

Permissible expansion of a 0.7-gravity natural gas without hydrate formation.

Discharge Head

$$H_d = \left\{ \left[(p_2 + p_{f2} + P_c) \times 2.31 \right] / \gamma \right\} + H_2$$

H_d	Discharge head of liquid being pumped
p_2	Discharge vessel operating pressure
p_{f2}	Pressure drop resulting from friction in the discharge piping
γ	Specific gravity of liquid, dimensionless
H_2	Operating or normal height of liquid in the discharge vessel above the pump reference
P_c	Discharge flow-control-valve losses

Calculating Total Dynamic Head (TDH)

$$H_{td} = H_d - H_s = \left\{ \left[(p_2 - p_1 + p_{f1} + p_{f2} + P_c) \times 2.31 \right] / \gamma \right\} + H_2 - H_1$$

154

H_{td} Total dynamic head
H_d Discharge head
H_s Suction head of liquid being pumped
p_2 Discharge vessel operating pressure
p_1 Suction-vessel operating pressure
p_{f1} Pressure drop resulting from friction in the suction pipe
p_{f2} Pressure drop resulting from friction in the discharge pipe
P_c Discharge flow-control-valve losses
γ Specific gravity of liquid, dimensionless
H_2 Operating or normal height of liquid in the discharge vessel above the pump reference
H_1 Height of liquid suction vessel above pump reference point

Power Requirements for Kinetic Energy Pumps

$$P_B = (q \times H_{td} \times \gamma) / 3{,}960e$$

Power Requirements for Positive-Displacement Pumps

$$P_B = (q \times \Delta p) / 1{,}714e$$

Energy Consumption

$$E / t = 17.9 P_B / e_m$$

P_B Brake horsepower, hp
e Pump efficiency factor obtained from the pump manufacturer
H_{td} Total dynamic head, ft
q Pump capacity, gal/min
γ Specific gravity of liquid, dimensionless
Δp Pressure difference, psi
E Electrical energy, kW-hour
t time, hr/D
e_m Motor efficiency, fraction

Centrifugal Pumps

Pump-Specific Speed

$$N_s = \frac{N\sqrt{q}}{\left(H_{td}'\right)^{0.75}}$$

N_s Pump-specific speed, dimensionless
N Pump-rotative speed, rev/min
q Pump capacity, gal/min
H_{td}' TDH per stage at the best efficiency point, ft

Constant Portion of System-Head Curves

$$H_{ws} = \left\{\left[(p_2 - p_1) \times 2.31\right] / \gamma\right\} + H_2 - H_1$$

H_{ws} Static difference between the suction and the discharge at zero flow, ft

Variable Portion of System-Head Curves

$$H_V = \left[\left(p_{f1} + p_{f2} + P_c\right) \times 2.31\right] / \gamma$$

H_V Head required to overcome friction as a result of flow, ft

Affinity Laws

$$q_2 = q_1\left(N_2 / N_1\right)$$

$$H_{td2} = H_{td1}\left(N_2 / N_1\right)^2$$

$$P_{B2} = P_{B1}\left(N_2 / N_1\right)^3$$

N_1 Old speed, rev/min
N_2 New speed, rev/min

$$q_2 = q_1\left(D_2 / D_1\right)$$

$$H_{td2} = H_{td1}\left(D_2 / D_1\right)^2$$

$$P_{B2} = P_{B1}\left(D_2 / D_1\right)^3$$

D_1 Old diameter, in.
D_2 New diameter, in.

$$q_2 = q_1\left(D_2 / D_1\right)\left(N_2 / N_1\right)$$

$$H_{td2} = H_{td1}\left(D_2 / D_1\right)^2\left(N_2 / N_1\right)^2$$

$$P_{B2} = P_{B1}\left(D_2 / D_1\right)^3\left(N_2 / N_1\right)^3$$

Positive-Displacement Pump

Cylinder Displacement of a Reciprocating Pump

$$s = \frac{A \times L_s \times N \times m}{231} \qquad \text{single-acting cylinder}$$

$$s = \frac{(2A - a) \times L_s \times N \times m}{231} \qquad \text{double-acting cylinder}$$

s Cylinder displacement, gal/min
A Plunger or piston area, in.^{2}
a Piston-rod cross-sectional area, in.^{2}
L_s Stroke length, in.
N Speed, rev/min
m Number of pistons or plungers

Pump Capacity

$$q = \frac{s(100 - S)}{100}$$

q Pump capacity, gal/min
s Cylinder displacement, gal/min
S Slip, %

5.4 Compressors

Isentropic (Adiabatic) Compression

$$P_1 V_1^k = P_2 V_2^k$$

Polytropic Compression

$$P_1 V_1^n = P_2 V_2^n$$

P_1 Specified absolute suction pressure, psia
V_1 Volume at pressure 1
P_2 Specified absolute discharge pressure, psia
V_2 Volume at pressure 2
k Ratio of specific heats, $= C_p / C_v$
n Polytropic exponent

Compressor Horsepower (GPSA 2004)

$$\text{Brake horsepower} = (22)\left(\frac{\text{ratio}}{\text{stage}}\right)(\text{number of stages})(\text{MMcfD})(F)$$

Where:
MMcfD = Compressor capacity referred to 14.4 psia and intake temperature
F = 1.0 for single-stage compression
      1.08 for two-stage compression
      1.10 for three-stage compression
For gases with a specific gravity in the 0.8 to 1.0 range, use a multiplication factor of 20 instead of 22.
For gases with compression ratios between 1.5 and 2.0, use a multiplication factor in the range of 16 to 18.

Isentropic (Adiabatic) Head

$$H_{is} = 53.3 z_{\text{avg}}\left(T_s / S\right)\left[k/(k-1)\right]\left[\left(P_d / P_s\right)^{(k-1)/k} - 1\right]$$

Polytropic Head

$$H_p = 53.3 z_{\text{avg}}\left(T_s / S\right)\left[k\eta_p/(k-1)\right]\left[\left(P_d / P_s\right)^{(k-1)/k\eta_p} - 1\right]$$

H_{is} Isentropic head, ft-lbf/lbm
H_p Polytropic head, ft-lbf/lbm
z_{avg} Average compressibility factor, dimensionless
T_s Suction temperature, °R
S Gas-specific gravity (standard atmospheric air = 1.00)
k Ratio of specific heats
P_d Discharge pressure, psia
P_s Suction pressure, psia
η_p Polytropic efficiency

Isentropic (Adiabatic) Efficiency

$$\eta_{is} = T_s\left[\left(P_d / P_s\right)^{(k-1)/k} - 1\right]/\left(T_d - T_s\right)$$

Polytropic Efficiency

$$\eta_p = \left[(k-1)/k\right]\ln\left(P_d / P_s\right)/\ln\left(T_d / T_s\right)$$

η_{is} Isentropic efficiency
η_p Polytropic efficiency
T_s Suction temperature, °R
T_d Discharge temperature (actual or predicted), °R
P_d Discharge pressure, psia

P_s Suction pressure, psia

k Ratio of specific heats

Compressibility Factor

$z = PV / nRT$

z Compressibility factor, dimensionless

P Pressure, psia

V Volume, ft³

T Temperature, °R

n Number of moles

R Constant for a specific gas, ft·lb/mol·°R

Actual (Inlet) Volume Flow

Actual cubic feet per minute (ACFM) = $WRT_s z_s / 144 P_s MW$

Actual cubic feet per minute = $19.6 T_s Q_g z_s / P_s$

W Mass flow, lbm/min

Q_g Standard volume flow, MMscf/D

R Universal gas constant, = 1,545

MW Molecular weight, lbm/mol

T_s Suction temperature, °R

P_s Absolute suction pressure, psia

z_s Compressibility at inlet

Power Requirement for Centrifugal Compressors

$GHP = WH_p / 33,000 \eta_p$

Power Requirement for Reciprocating Compressors

$$GHP = \left[(0.004367)(P_1 / V_1)(k / k - 1)(P_2 / P_1^{k-1/k} - 1) \right] / CE$$

GHP Gas horsepower, hp

W Mass flow, lbm/min

H_p Polytropic head, ft-lbf/lbm

η_p Polytropic efficiency

P_1 Inlet pressure, psia

V_1 Inlet volume, ACFM

P_2 Discharge pressure, psia

CE Compression efficiency, assume 0.85 for estimating purposes

Discharge Temperature for Compression

$$\ln(T_2 / T_1) = \left(\left\{ \left[(k-1) / k \right] \left[\ln(P_2 / P_1) \right] \right\} \right) / \eta_p$$

T_2 Estimated absolute discharge temperature, °R

T_1 Specified absolute suction temperature, °R

P_1 Specified absolute suction pressure, psia

P_2 Specified absolute discharge pressure, psia

k Ratio of specific heats

η_p Assumed polytropic efficiency

    $\approx$ 0.72 to 0.85 for centrifugal compressors

    $\approx$ 1.00 for reciprocating compressors

Compression Ratio per Section

$$R_{\text{sect}} = \left(P_2 / P_1\right)^{1/n}$$

R_{sect} Compression ratio per section
n Number of sections

Reciprocating Compressors

Compressor Capacity

$$q_a = E_v\,(PD)$$

$$q_g = 35.4\left[\left(q_a P_s\right)/\left(T_s z_s\right)\right]$$

$$Q_g = 0.051\left[\left(q_a P_s\right)/\left(T_s z_s\right)\right]$$

q_a Inlet capacity of the cylinder at actual inlet conditions, ACFM
E_v Volumetric efficiency
PD Piston displacement, ACFM
q_g Inlet capacity of the cylinder, scf/min
Q_g Inlet capacity of the cylinder, MMscf/D
T_s Suction temperature, °R
P_s Absolute suction pressure, psia
z_s Compressibility at inlet

Piston Displacement for Single-Acting Cylinder (Head-End Displacement)

$$PD = \left(d_c^2 SN\right)/2{,}200$$

Piston Displacement for Single-Acting Cylinder (Crank-End Displacement)

$$PD = \left\lfloor\left(d_c^2 - d_r^2\right)SN\right\rfloor/2{,}200$$

Piston Displacement for Double-Acting Cylinder

$$PD = \left\lfloor\left(2d_c^2 - d_r^2\right)SN\right\rfloor/2{,}200$$

PD Piston displacement, ACFM
S Stroke, in.
N Compressor speed, rpm
d_c Cylinder diameter, in.
d_r Rod diameter, in.

Clearance Volume for Single-Acting Cylinder (Head-End Clearance)

$$\%C = 127\left[C_{HE}/\left(d_c^2 S\right)\right]$$

Clearance Volume for Single-Acting Cylinder (Crank-End Clearance)

$$\%C = 127\left\{C_{HE}/\left[\left(d_c^2 - d_r^2\right)S\right]\right\}$$

Clearance Volume for Double-Acting Cylinder (Head-End and Crank-End Clearance)

$$\%C = 127\left\{\left(C_{HE} + C_{CE}\right)/\left[d_c^2 + \left(d_c^2 - d_r^2\right)S\right]\right\}$$

$\%C$ Cylinder clearance, %
C_{HE} Head-end clearance, in.^{3}
C_{CE} Crank-end clearance, in.^{3}

d_c Cylinder ID, in.
d_r Rod diameter, in.
S Stroke length, in.

Volumetric Efficiency (Approximate)

$$E_v = 96 - R - C\left\{\left[\left(R^{1/k}\right)\left(\frac{z_s^2}{z_d^2}\right)\right]-1\right\}-L$$

E_v Volumetric efficiency
R Compression ratio
C Cylinder clearance, % of piston-swept volume
z_s Inlet compressibility factor
z_d Discharge compressibility factor
k Ratio of specific heats, C_p/C_v
L Slippage of gas past piston rings, % [1% for high-speed separable, 4% for valve losses, 5% for nonlubricated compressors, and 4% for heavy gases (propane, CO_2, etc) service]

Rod Load for Single-Acting Cylinder (Head-End)

$$RL_c = a_p\left(P_d - P_u\right)+a_r P_u$$

$$RL_t = a_p\left(P_u - P_s\right)-a_r P_u$$

Rod Load for Single-Acting Cylinder (Crank-End)

$$RL_c = a_p\left(P_u - P_s\right)+a_r P_s$$

$$RL_t = a_p\left(P_d - P_u\right)-a_r P_d$$

Rod Load for Double-Acting Cylinder

$$RL_c = a_p\left(P_d - P_s\right)+a_r P_s$$

$$RL_t = a_p\left(P_d - P_s\right)-a_r P_d$$

RL_c Rod load-in compression, lbf
RL_t Rod load-in tension, lbf
a_p Cross-section area of piston, in.^{2}
a_r Cross-section area of rod, in.^{2}
P_d Discharge pressure, psia
P_s Suction pressure, psia
P_u Pressure in unloaded end, psia

Piping Vibration

$$f_p = \left(N/60\right)^n$$

f_p Compressor pulsation frequency, cycles/sec
N Compressor speed, rpm
n Cylinder factor, = 1 for single-acting cylinder, = 2 for double-acting cylinder

5.5 Pipelines

Pressure Drop Equations

Bernoulli Equation

$$Z_1 + \frac{144P_1}{\rho_1} + \frac{V_1^2}{2g} = Z_2 + \frac{144P_2}{\rho_2} + \frac{V_2^2}{2g} + H_L$$

$$H_L = \frac{fLV^2}{2Dg} \qquad \text{from Darcy's}$$

$$\Delta P = 0.0013 \frac{f\rho LV^2}{d} \qquad \text{from Darcy's}$$

Z Elevation head, ft
P Pressure, psi
ρ Density, lbm/ft³
V Velocity, ft/sec
g Gravitational constant, ft/sec²
H_L Head loss, ft
f Moody friction factor, dimensionless
L Pipe length, ft
D Pipe diameter, ft
ΔP Pressure drop, psi
d Pipe inside diameter, in.

Reynolds Number for Liquids

$$\text{Re} = \frac{92.1(S_G)Q_L}{d\mu}$$

Reynolds Number for Gases

$$\text{Re} = \frac{20,100 S Q_g}{d\mu}$$

μ Viscosity, cp
d Pipe inside diameter, in.
S_G Specific gravity of gas relative to water, (water = 1)
S Specific gravity of gas
Q_L Liquid flow rate, B/D
Q_g Gas flow rate, MMscf/D

Pressure Drop for Liquid Flow

$$d^5 = \left(11.5 \times 10^{-6}\right) \frac{fLQ_L^2 (SG)}{\Delta P}$$

d Pipe ID, in.
f Moody friction factor, dimensionless
L Pipe length, ft
SG Specific gravity of liquid relative to water
Q_L Liquid flow rate, B/D
ΔP Pressure drop, psi (total pressure drop)

Hazen-Williams Equation (Water in Turbulent Flow at 60°F)

$$H_L = 0.00208 \left(\frac{100}{C}\right)^{1.85} \left(\frac{\text{gpm}}{d^{2.63}}\right)^{1.85} \qquad L = 0.015 \left(\frac{Q_L^{1.85} L}{d^{4.87} C^{1.85}}\right) \qquad \Delta P = 0.43\, H_L$$

H_L Head loss because of friction, ft
L Pipe length, ft
C Friction-factor constant, dimensionless
d Pipe ID, in.
Q_L Liquid flow rate, B/D
gpm   Liquid flow rate, gal/min
ΔP Pressure drop, psi

Pressure Drop for Gas Flow (General)

$$w^2 = \left| \frac{144\,gA^2}{V_1'\left(\dfrac{fL}{D} + 2\log_e \dfrac{P_1}{P_2}\right)} \right| \times \left(\frac{P_1^2 - P_2^2}{P_1}\right)$$

Simplified Practical Equation

$$P_1^2 - P_2^2 = 25.2\left(\frac{SQ_g^2 ZTfL}{D^5}\right)$$

w Rate of flow, lbm/sec
g Acceleration of gravity, 32.2 ft/sec²
A Cross-sectional area of pipe, ft²
V_1' Specific volume of gas at upstream conditions, ft³/lbm
f Friction factor, dimensionless
L Length, ft
D Pipe ID, in.
P_1 Upstream pressure, psia
P_2 Downstream pressure, psia
S Specific gravity of gas
Q_g Gas flow rate, MMscf/D
Z Gas compressibility factor, dimensionless
T Flowing temperature, °R

Weymouth Equation

$$Q_g = 1.1 d^{2.67}\left(\frac{P_1^2 - P_2^2}{LSZT_1}\right)^{1/2}$$

Q_g Gas flow rate, MMscf/D
d Pipe ID, in.
P_1 Upstream pressure, psia
P_2 Downstream pressure, psia
L Length, ft
Z Gas compressibility factor, dimensionless
T_1 Temperature of the gas at inlet, °R
S Specific gravity of gas

Panhandle Equation

$$Q_g = 0.028 E\left(\frac{P_1^2 - P_2^2}{L_m S^{0.961} ZT_1}\right)^{0.51} d^{2.53}$$

Q_g Gas flow rate, MMscf/D
d Pipe ID, in.
P_1 Upstream pressure, psia
P_2 Downstream pressure, psia
L_m Length, miles
Z Gas compressibility factor, dimensionless
T_1 Temperature of the gas at inlet, °R
S Specific gravity of gas
E Efficiency factor (new pipe: 1.0; good operating conditions: 0.95; average operating conditions: 0.85)

Spitzglass Equation

$$Q_g = 0.09\left[\frac{\Delta h_w d^5}{SL(1 + 3.6/d + 0.03d)}\right]^{1/2}$$

Assumptions: $f = (1+3.6/d+0.03d)(1/100)$, $T = 520$ °R, $P_1 = 15$ psia, $z = 1.0$, $\Delta P \leq 10\%$ P_1

Q_g Gas flow rate, MMscf/D
d Pipe ID, in.
Δh_w Pressure loss, inches of water
L Length, ft
S Specific gravity of gas

Two-Phase Pressure Drop

$$\Delta P = \frac{3.4 \times 10^{-6} \, fLw^2}{\rho_M d^5}$$

ΔP Friction pressure drop, psi
f Friction factor, dimensionless
L Length, ft
w Rate of flow of mixture, lbm/hr
ρ_M Density of the mixture, lbm/ft³
d Pipe ID, in.

$$W = 3{,}180 Q_g S + 14.6 Q_L (S_G)$$

W Rate of flow of mixture
Q_g Gas flow rate, MMscf/D
Q_L Liquid flow rate, B/D
S Specific gravity of gas at standard conditions, lbm/ft³ (air = 1)
S_G Specific gravity of liquid, relative to water, lbm/ft³

$$\rho_M = \frac{12{,}409 \, (S_G) P + 2.7 RSP}{198.7 P + RTZ}$$

P Operating pressure, psia
R Gas/liquid ratio, ft³/bbl
T Operating temperature, °R
S Specific gravity of gas at standard conditions, lbm/ft³ (air = 1)
S_G Specific gravity of liquid, relative to water, lbm/ft³
Z Gas compressibility factor, dimensionless

Pressure Drop Due to Changes in Elevation

$$\Delta P_Z \approx 0.433 (SG) \Delta Z$$

ΔP_Z Pressure drop because of elevation increase in the segments, psi
SG Specific gravity of liquid in the segment, relative to water, lbm/ft³
ΔZ Increase in elevation for segment, ft

Pressure Drop Caused by Valves and Fittings

Resistance Coefficient

$$H_L = K_r \frac{V^2}{2g}$$

H_L Head loss, ft
K_r Resistance coefficient, dimensionless
V Velocity, ft/sec
g Gravity acceleration, ft/sec²

Flow Coefficients

$$C_V = \frac{29.9 d^2}{K_r^{1/2}}$$

$$\Delta P = 8.5 \times 10^{-4} \left(\frac{Q_L}{C_V} \right)^2 (SG)$$

ΔP Pressure drop, psi
C_V Flow coefficient for liquid
Q_L Liquid flow rate, B/D
SG Liquid-specific gravity relative to water
d Pipe ID, in.
K_r Resistance coefficient, dimensionless

Equivalent Lengths (L_e)

$$L_e = \frac{K_r D}{f} \qquad L_e = \frac{K_r D}{12 f} \qquad L_e = \frac{74.5 d^5}{f(C_V)^2}$$

K_r Resistance coefficient, dimensionless
d Pipe inside diameter, in.
D Diameter of the pipe, ft
f Moody friction factor, dimensionless
C_V Flow coefficient for liquids

Flow Through a Choke

Single-Phase Gas Flow

$$q_g = \frac{27.611 C_d p_{wh} d^2 T_{sc}}{p_{sc} \sqrt{\gamma_g T_{wh} Z}} \sqrt{\left(\frac{k}{k-1} \right) \left[y^{2/k} - y^{(k+1)/k} \right]}$$

Two-Phase Critical Flow

$$p_{wh} = \frac{A_1 q_L R^{A_2}}{d^{A_3}}$$

A_{1-3} Coefficient from table
C_d Discharge coefficient, dimensionless
d Pipe ID, in.
k Specific heat-capacity ratio, C_p / C_v, dimensionless
p_{sc} Standard pressure, m/Lt², psia
p_{wh} Wellhead pressure, m/Lt², psia
q_g Gas flow rate, L³/t, Mscf/D
q_L Liquid flow rate, L³/t, STB/D
T_{sc} Standard temperature, T, °R
T_{wh} Wellhead temperature, T, °R
R Gas-liquid ratio, scf/STB
y Ratio of downstream pressure to upstream pressure, p_1 / p_2, dimensionless
Z Gas compressibility factor, dimensionless
γ_g Gas-specific gravity, dimensionless

Researcher	A_1	A_2	A_3
Gilbert	3.86×10^{-3}	0.546	1.89
Ros	4.26×10^{-3}	0.500	2.00
Baxendell	3.12×10^{-3}	0.546	1.93
Achong	1.54×10^{-3}	0.650	1.88

Table 5.2—Coefficient values for A_1, A_2, and A_3.

Pipe Wall Thickness

Minimum Internal Yield Pressure (MIYP): Barlow Equation

The MIYP of the pipe is determined by the internal yield pressure formula:

$$P_B + \left(\frac{2Y_p t}{D} \right)$$

P_B Minimum burst pressure, psi
Y_p Minimum yield strength, psi
t Nominal wall thickness, in.
D Nominal outside pipe diameter

This equation, commonly known as the Barlow equation, calculates the internal pressure at which the tangential (or hoop) stress at the inner wall of the pipe reaches yield strength (YS) of the material.

Basic Formula for Thin-Wall Cylinder

$$t = \frac{P d_o}{2(H_s + P)}$$

H_s Hoop stress in pipe wall, psi
t Pipe wall thickness, in.
P Internal pressure of the pipe, psi
d_o Pipe OD, in.

Wall Thickness Calculations — Using ASME B31.3 Code

$$t = t_e + t_{th} + \left[\frac{P d_o}{2(SE + PY)} \right] \left(\frac{100}{100 - T_{ol}} \right)$$

t Minimum-design wall thickness, in.
t_e Corrosion allowance, in.
t_{th} Thread or groove depth, in.
P Allowable internal pressure in pipe, psi
d_o Pipe OD, in.
S Allowable stress for pipe, psi
E Longitudinal weld-joint factor [1.0 seamless, 0.95 electric fusion weld, double butt, straight or spiral seam API 5L, 0.85 electric resistance weld (ERW), 0.6 furnace butt weld]
Y Derating factor (0.4 for ferrous materials operating below 900°F)
T_{ol} Manufacturers allowable tolerance, % (12.5 pipe up to 20-in. OD, 10 pipe > 20-in. OD, API 5L)

*Wall Thickness Calculations — Using ASME B31.4 Code (see **Table 5.3**)*

$$t = \frac{P d_o}{2(FES_Y)}$$

t Minimum-design wall thickness, in.
P Internal pressure in pipe, psi
d_o Pipe OD, in.
S_Y Minimum-yield stress for pipe, psi (see Table 9.11 in Appendix)
F Designing factor, 0.72 for all locations
E Longitudinal weld-joint factor [1.0 seamless, ERW, double-submerged-arc weld and flash weld; 0.80 electric fusion (arc) weld and electric fusion weld, 0.60 furnace butt weld]

Wall Thickness Calculations — Using ASME B31.8 Code

$$t = \frac{P d_o}{2(FETS_Y)}$$

Table 5.3—ANSI Pipe Schedules
(Courtesy of Natural Gas Processors Suppliers Assn.)

Upper figures indicate wall thickness in in.   Lower figures indicate weight per foot in lbm

Pipe Size	O.D in in.	5	10	20	30	40	STD.	60	80	XH	100	120	140	160	XXH
⅛	.405	.035 .1383	.049 .1863			.068 .2447	.068 .2447		.095 .3145	.095 .3145					
¼	.540	.049 .2570	.065 .3297			.088 .4248	.088 .4249		.119 .5351						
⅜	.675	.049 .3276	.065 .4235			.091 .5676	.091 .5676		.126 .7388	.126 .7388					
½	.840	.065 .5383	.083 .6710			.109 .8510	.109 .8510		.147 1.088	.147 1.088				.187 1.304	.294 1.714
¾	1.050	.065 .6838	.083 .8572			.113 1.131	.113 1.131		.154 1.474	.154 1.474				.218 1.937	.308 2.441
1	1.315	.065 .8678	.109 1.404			.133 1.679	.133 1.679		.179 2.172	.179 2.172				.250 2.844	.358 3.659
1¼	1.660	.065 1.107	.109 1.806			.140 2.273	.140 2.273		.191 2.997	.191 2.997				.250 3.765	.382 5.214
1½	1.900	.065 1.274	.109 2.085			.145 2.718	.145 2.718		.200 3.631	.200 3.631				.281 4.859	.400 6.408
2	2.375	.065 1.604	.109 2.638			.154 3.653	.154 3.653		.218 5.022	.218 5.022				.343 7.444	.436 9.029
2½	2.875	.083 2.475	.120 3.531			.203 5.793	.203 5.793		.276 7.661	.276 7.661				.375 10.01	.552 13.70
3	3.500	.083 3.029	.120 4.332			.216 7.576	.216 7.576		.300 10.25	.300 10.25				.437 14.32	.600 18.58
3½	4.0	.083 3.472	.120 4.973			.226 9.109	.226 9.109		.318 12.51	.318 12.51					.636 22.85
4	4.50	.083 3.915	.120 5.613			.237 10.79	.237 10.79	.281 12.66	.337 14.98	.337 14.98		.437 19.01		.531 22.51	.674 27.54
4½	5.0					.247 12.53				.355 17.61					.710 32.53
5	5.563	.109 6.349	.134 7.770			.258 14.62	.258 14.62		.375 20.78	.375 20.78		.500 27.04		.625 32.96	.750 38.55
6	6.625	.109 7.585	.134 9.289			.280 18.97	.280 18.97		.432 28.57	.432 28.57		.562 36.39		.718 45.30	.864 53.16
7	7.625						.301 23.57			.500 38.05					.875 63.08
8	8.625	.109 9.914	.148 13.40	.250 22.36	.277 24.70	.322 28.55	.322 28.55	.406 35.64	.500 35.64	.500 43.39	.593 50.87	.718 60.63	.812 67.76	.906 74.69	.875 72.42
9	9.625						.342 33.90			.500 48.72					
10	10.75	.134 15.19	.165 18.70	.250 28.04	.307 34.24	.365 40.48	.365 40.48	.500 54.74	.593 64.33	.500 54.74	.718 76.93	.843 89.20	1.000 104.1	1.125 115.7	
11	11.75						.375 45.55			.500 60.07					
12	12.75	.165 22.18	.180 24.20	.250 33.38	.330 43.77	.406 53.53	.375 49.56	.562 73.16	.687 88.51	.500 65.42	.843 107.2	1.000 125.5	1.125 139.7	1.312 160.3	
14	14.0		.250 36.71	.312 45.68	.375 54.57	.437 63.37	.375 54.57	.593 84.91	.750 106.1	.500 72.09	.937 130.7	1.093 150.7	1.250 170.2	1.406 189.1	
16	16.0		.250 42.05	.312 52.36	.375 62.58	.500 82.77	.375 62.58	.656 107.5	.843 136.5	.500 82.77	1.031 164.8	1.218 192.3	1.437 223.5	1.593 245.1	
18	18.0		.250 47.39	.312 59.03	.437 82.06	.562 104.8	.375 70.59	.750 138.2	.937 170.8	.500 93.45	1.156 208.0	1.375 244.1	1.562 274.2	1.781 308.5	
20	20.0		.250 52.73	.375 78.60	.500 104.1	.593 122.9	.375 78.60	.812 166.4	1.031 208.9	.500 104.1	1.280 256.1	1.500 296.4	1.750 341.1	1.968 379.0	
22	22.0		.250 58.07	.375 86.61	.500 114.8		.375 86.61	.875 197.4	1.125 250.8	.500 114.8	1.375 302.9	1.625 353.6	1.875 403.0	2.125 451.1	
24	24.0		.250 63.41	.375 94.62	.562 104.8	.687 171.2	.375 94.62	.968 238.1	1.218 296.4	.500 125.5	1.531 367.4	1.812 429.4	2.062 483.1	2.343 541.9	
26	26.0		.312 85.60	.500 136.2			.375 102.6			.500 136.2					
28	28.0		.312 92.26	.500 146.8	.625 182.7		.375 110.6			.500 146.8					
30	30.0		.312 98.93	.500 157.5	.625 196.1		.375 118.6			.500 157.5					
32	32.0		.312 105.6	.500 168.2	.625 209.4	.688 230.1	.375 126.7			.500 168.2					
34	34.0		.344 123.7	.500 178.9	.625 222.8	.688 244.8	.375 134.7			.500 178.9					
36	36.0		.312 118.9	.500 189.6	.625 236.1	.750 282.3	.375 142.7			.500 189.6					
42	42.0						.375 166.7			.500 221.6					
48	48.0						.375 190.7			.500 253.6					

Minimum-design wall thickness, in.

P Internal pressure in pipe, psi

d_o Pipe OD, in.

S_Y Minimum-yield stress for pipe, psi (see Table 9.12 in Appendix)

F Designing factor (see Table 9.13 in Appendix)

E Longitudinal weld-joint factor (see Table 9.14 in Appendix)

T Temperature-derating factor (see Table 9.15 in Appendix)

For additional information on Wall Thickness Calculations see Appendix.

Velocity Calculations

Liquid-Line Sizing

$$V = 0.012 \frac{Q_L}{d^2}$$

V Flow velocity, ft/sec

Q_L Fluid flow rate, B/D

d Pipe ID, in.

Gas-Line Sizing

$$V_g = 60 \frac{Q_g T z}{d^2 P}$$

V_g Gas velocity, ft/sec

Q_g Gas flow rate, MMscf/D

T Gas flowing temperature, °R

P Flowing pressure, psia

z Compressibility factor, dimensionless

d Pipe ID, in.

Multiphase-Line Sizing

$$V_e = \frac{C}{\rho_M^{1/2}}$$

$$\rho_M = \frac{(12,409)(S_G)P + (2.7)RSP}{(198.7)P + zRT}$$

V_e Erosional velocity of the mixture, ft/sec

C Empirical constant

ρ_M Average density of the mixture at flowing conditions, lbm/ft³

S_G Liquid-specific gravity relative to water

S Specific gravity of the gas relative to air

T Gas/liquid flowing temperature, °R

P Flowing pressure, psia

z Compressibility factor, dimensionless

R Gas/liquid ratio, ft³/bbl

Determining the Pipe Size

$$d = \left[\frac{\left(11.9 + \frac{zTR}{16.7P} \right) Q_L}{1,000V} \right]^{1/2}$$

d Pipe ID, in.

z Compressibility factor, dimensionless

R Gas/liquid ratio, ft³/bbl

T Gas/liquid flowing temperature, °R
P Flowing pressure, psia
V Maximum allowable velocity, ft/sec
Q_L Liquid flow rate, B/D

Pipeline Design

Minimum Bending Radius of Concrete-Coated Pipe

$$R = EC / S_B$$

R Bending radius, in.
E Modulus of elasticity for concrete = 3,000,000 psi
C Pipe radius + enamel thickness + concrete thickness, in.
S_B 2,500 psi

Minimum Bending Radius of Steel Pipe

$$R = EC / fS_Y - PD / 4t$$

R Bending radius, in.
P Design pressure, psi
D Pipe OD, in.
t Pipe wall thickness, in.
E Modulus of elasticity for concrete = 30,000,000 psi
C Pipe radius, in.
S_Y Pipe-specified minimum-yield strength, psi
f Stress factor: use 75 to 85% for offshore design

Pipeline Pigging

Slug Catchers

$$Q_L = 5,000 Q_g Tz / P$$

Q_L Liquid flow rate in front of the pig, B/D
Q_g Gas flow rate behind the pig, MMscf/D
T Temperature, °R
P Line pressure, psi
z Compressibility factor, dimensionless

$$(\text{Vol})_{SC} = (\text{Vol}) - Q_d T_R$$

$(\text{Vol})_{SC}$ Volume of slug catcher, bbl
Vol   Volume of liquid holdup, bbl
Q_d Design-liquid dump rate from the slug catcher, B/D
T_R Time during which slug is processed, days, = Vol/Q_L
Q_L Liquid flow rate in front of the pig, B/D

Class	−20 to 100°F	200°F
150	285	260
300	740	675
400	990	900
600	1,480	1,350
900	2,220	2,025
1,500	3,705	3,375
2,500	6,170	5,625

Pressure ratings for temperatures between 100 and 200°F should be interpolated between the values of 100 and 200°F.

Table 5.4—Pressure ratings vs. temperature for materials group 1.1-27, class 150, 300, 400, 600, 900, 1,500, and 2,500 (Courtesy of ANSI/ASME).

Table G1—Temperature Ratings

Classification	Operating Range (°F)
X	0 to 350
Y	0 to 650

Table G2—Pressure/Temperature Ratings

	Temperature (°F)								
	0 to 250	300	350	400	450	500	550	600	650
Rated	2,000	1,955	1,905	1,860	1,810	1,735	1,635	1,540	1,430
Working	3,000	2,930	2,860	2,785	2,715	2,605	2,455	2,310	2,145
Pressure (psi)	5,000	4,880	4,765	4,645	4,525	4,340	4,090	3,850	3,575

Table 5.5—Pressure ratings vs. temperature for API flanges (Courtesy of API).

5.6 Flow Measurements Units

Differential Pressure Flowmeters

$$q_v = \frac{C}{\sqrt{1-\beta^4}} \cdot \varepsilon \cdot \frac{\pi}{4} d^2 \frac{\sqrt{2\rho_1 \Delta p}}{\rho}$$

q_v Volume flow rate, scf/hr
C Coefficient of discharge
β Diameter ratio, $= d/D$
d Orifice diameter, in.
D Pipe ID, in.
ε Expansion factor (for compressible media only)
d Inside diameter of the orifice plate, in.
ρ_1 Density at the upstream pressure-tapping cross section
ρ Density of the fluid at the appropriate conditions of pressure and temperature
Δp Differential pressure

Variable Area Flowmeters

$$q_v = \frac{\alpha}{\rho_m} D_s \sqrt{gm_s \rho_m \left(1 - \rho_m / \rho_s\right)}$$

q_v Volume flow rate, scf/hr
ρ_m Density of the measuring medium
α Flow coefficient, $= \sqrt{1/c_w}$
c_w Resistance coefficient
D_s Diameter of the float at the reading edge, ft

m_s Mass of the float, lbm
ρ_m Density of the measuring medium, lbm/ft³
ρ_s Density of the float, lbm/ft³
g Gravitational acceleration

Electromagnetic Flowmeters

$$q_v = K\frac{U_0}{U_{\text{Ref}}}$$

q_v Volume flow rate, scf/hr
K Calibration factor
U_0 Signal voltage
U_{Ref} Voltage generated in the sensor/transmitter

Ultrasonic Flowmeters

$$q_{adm} = q_b + q_t + \sum_{i=1}^{n}\left\{\frac{\overline{v}_{ai} + \overline{v}_{ai+1}}{2}\left[A(z_{i+1}) - A(z_i)\right]\right\}$$

q_{adm} Total discharge
q_b Flow rate in the bottom section, with the bottom velocity obtained from the lowest path velocity by correction for bottom friction
q_t Flow rate at the highest active section with velocity v_{top} interpolated from the velocity profile
$\overline{v}_{ai}$ Mean velocity along the ith acoustic path
$A(z_i)$ Cross section below the ith path
v_{top} Velocity at highest active section interpolated from the velocity profile

Vortex-Shedding Flowmeters

$$q_v = \left(\frac{\rho_f}{\rho_b}\right) \times \frac{f}{K}$$

q_v Volume flow rate, scf/hr
ρ_f Fluid density at flowing conditions
ρ_b Base density
f Vortex-shedding frequency
K K factor

Thermal-Mass Flowmeters

$$q_m = \frac{(P - L) \times C}{c_p \times \Delta T}$$

q_m Mass flow rate, lbm/sec
P Constant electrical heating power
L Thermal power dissipation
C Device constant
ΔT Resultant temperature difference
c_p Heat capacity

Basic Orifice Meter

$$q_m = C_d E_v Y (\pi / 4) d^2 \left(2g_c \rho_{t,p} \Delta P\right)^{0.5}$$

q_m Mass flow rate, lbm/sec
C_d Orifice plate coefficient of discharge, dimensionless
E_v Velocity of approach factor, dimensionless
Y Expansion factor, dimensionless
d Orifice plate bore diameter calculated at flowing temperature, T_f, ft

g_c Dimensional conversion constant, lbm-ft/lbf-sec²
$\rho_{t,p}$ Density of the fluid at flowing conditions $\left(P_f, T_f\right)$, lbm/ft³
P_f Pressure of fluid
T_f Temperature of fluid
ΔP Orifice differential pressure, lbf/in.²

Note: An orifice meter installed backwards will read $\approx$15% low.

Orifice Meter Equation for Natural Gas

$$q_v = 218.573 C_d E_v Y_l d^2 \left(T_b / P_b\right) \sqrt{\frac{P_{fl} Z_{bair} h_w}{G_r Z_{fl} T_f}}$$

q_v Volume flow rate, scf/hr
C_d Orifice plate coefficient of discharge, dimensionless
E_v Velocity of approach factor, dimensionless
Y_l Expansion factor (upstream tap), dimensionless
d Orifice plate bore diameter calculated at flowing temperature, T_f, in.
G_r Real-gas relative density (specific gravity)
h_w Orifice differential pressure of water at 60°F, in.
P_b Base pressure, psia
P_{fl} Flowing pressure (upstream tap), psia
T_b Base temperature, °R
T_f Flowing temperature, °R
Z_{bair} Compressibility at base conditions $\left(P_b, T_b\right)$
Z_{fl} Compressibility (upstream flowing conditions P_{fl}, T_f)

Transit-Time Ultrasonic Meters

$$q_v = V_m \left(\frac{\pi D^2}{4}\right) = K \left(\frac{t_{du} - t_{ud}}{t_{du} \times t_{ud}}\right)$$

q_v Volume flow rate, scf/hr
t_{du} Transit time from transducer D to U, $= L/\left(C - V_m \cos\theta\right)$
t_{ud} Transit time from transducer U to D, $= L/\left(C + V_m \cos\theta\right)$
L Distance between transducers U and D (path length)
C Velocity of sound in the gas
V_m Mean velocity of the flowing gas
θ Angle between acoustic path and meter axis
D Diameter of the meter bore, ft
K Constant for a specific meter application

Turbine Meter Equation for Gas

$$q_b = q_f M_f \left(\frac{P_f T_b Z_b}{P_b T_f Z_f}\right)$$

q_b Flow rate at base conditions
q_f Flow rate at operating conditions (meter reading)
M_f Meter factor to correct meter output based on calibration
P_f Pressure-flowing conditions
P_b Base pressure, set by agreement near atmospheric pressure, °R
T_b Base temperature, set by agreement at 60°F
T_f Flowing temperature, °R
Z_b Compressibility at base pressure and temperature
Z_f Compressibility at flowing pressure and temperature

Flow-Calculation Guide Equations

Gas → Turbine, Displacement or Ultrasonic

- Mass: $lb/hr = \rho \cdot MF \cdot IV$ — Eq 3-1
- Volume: $scf/hr = F_{pb} \cdot F_{tm} \cdot F_{pm} \cdot S \cdot MF \cdot IV \cdot F_{tb}$ — Eq 3-2

Gas → Orifice

- Mass: $lb/hr = \sqrt{h_w} \cdot F_n (F_c + F_{sl}) \cdot \sqrt{\rho} \cdot 1.0618 \cdot F_a \cdot Y$ — Eq 3-3
- Volume: $scf/hr = \sqrt{h_w} \cdot \sqrt{P_f} \cdot F_n (F_c + F_{sl}) \cdot F_{pb} \cdot F_{tf} \cdot F_{pv} \cdot F_g \cdot Y \cdot F_{tb} \cdot F_a$ — Eq 3-4

Gas → Coriolis

- Mass: $lb/hr = MF \cdot IM$ — Eq 3-8
- Volume: $scf/hr = MF \cdot IM / \rho_b$ — Eq 3-9

Liquid → Turbine, Displacement or Ultrasonic

- Mass: $lb/hr = \rho \cdot MF \cdot IV$ — Eq 3-1
- Volume: $gal/hr = C_{tl} \cdot C_{pl} \cdot MF \cdot IV$ — Eq 3-5

Liquid → Orifice

- Mass: $lb/hr = \sqrt{h_w} \cdot F_n (F_c + F_{sl}) \cdot \sqrt{\rho} \cdot 1.0618 \cdot F_a$ — Eq 3-6
- Volume: $gal/hr = \sqrt{h_w} \cdot F_n (F_c + F_{sl}) \cdot F_{gt} \cdot F_a$ — Eq 3-7

Liquid → Coriolis

- Mass: $lb/hr = MF \cdot IM$ — Eq 3-8
- Volume: $gal/hr = C_{tl} \cdot C_{pl} \cdot MF \cdot IV$ — Eq 3-10

Steam → Orifice

- Mass: $lb/hr = \sqrt{h_w} \cdot F_n (F_c + F_{sl}) \cdot F_a \cdot Y \cdot F_a$ — Eq 3-11

Steam → Vortex

- Mass: $lb/hr = \rho \cdot IV$ — Eq 3-12

Other Equations

$$F_{pb} = \frac{14.73}{P_b}$$

$$F_{tb} = \frac{460 + T_b}{520}$$

$$F_{tf} = \sqrt{\frac{520}{460 + T_f}}$$

$$F_g = \sqrt{\frac{1.0000}{G}} \qquad F_{gt} = \frac{1.0057}{\sqrt{G_l}} \sqrt{\frac{G_f}{G_l}}$$

(See Fig. 23-10 and 23-11 for G_f/G_l conversions)

$$Q = C' \sqrt{h_w \cdot P_f}$$

$$C' = F_n (F_c + F_{sl}) \cdot F_{pb} \cdot F_{tf} \cdot F_{pv} \cdot F_g \cdot Y \cdot F_{tb} \cdot F_a$$

$$F_a = 1 + [0.0000185 (T_f - T_{meas})] \text{ for 304/316 Stainless Steel}$$

$$C_{pl} = \frac{1}{1 - (F \cdot Dp)}$$

$$F_n (F_c + F_{sl}) = \text{Key} = \text{orifice factor (Fig. 3-13)}$$

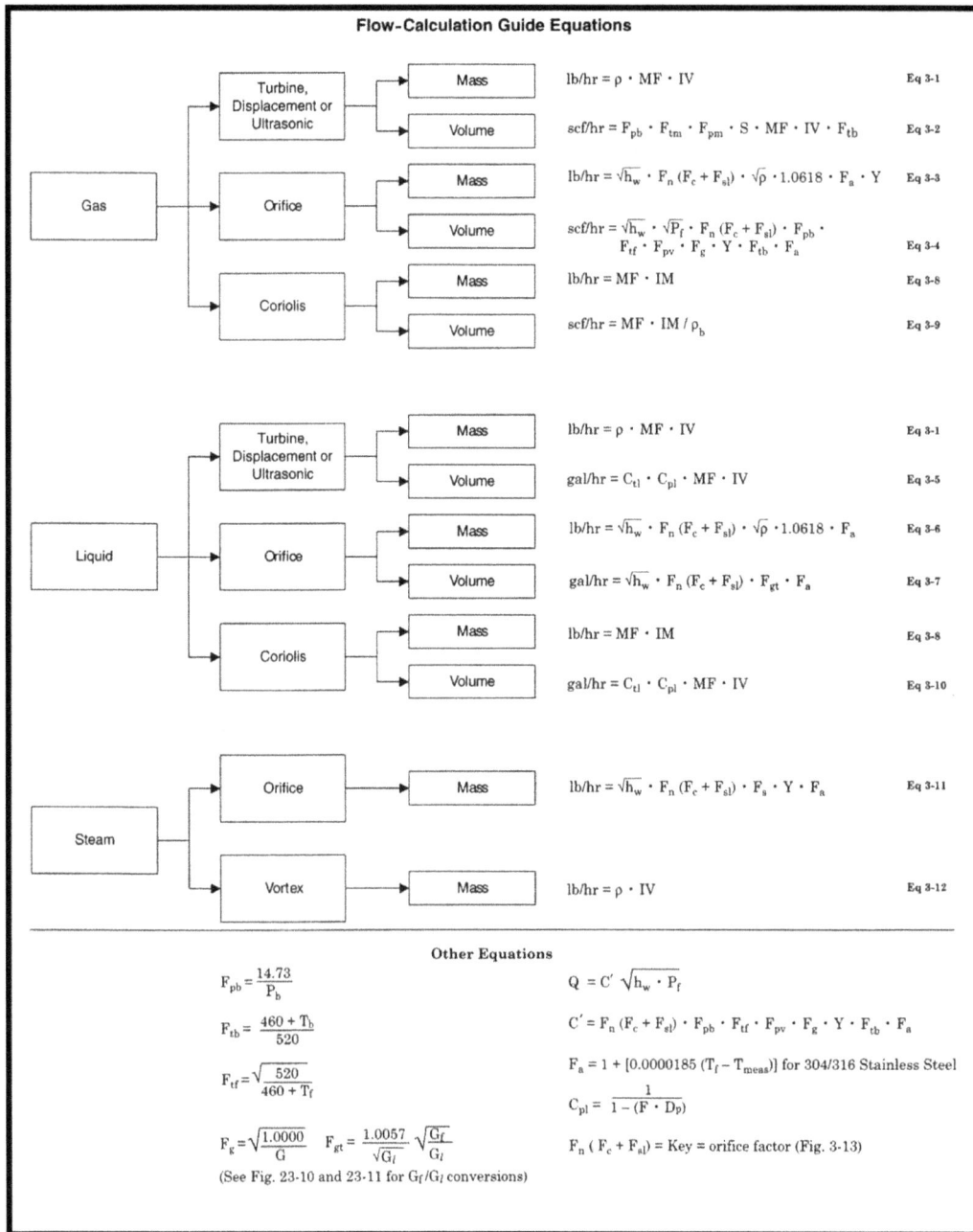

Fig. 5.1—Flow-calculation guide equations (GPSA 2012).

Nomenclature

C = Pitot tube flow coefficient

C' = the product of multiplying all orifice correction factors

C_{pl} = liquid pressure correction factor. Correction for the change in volume resulting from application of pressure. Proportional to the liquid compressibility factor, which depends upon both relative density and temperature.

C_{ps} = correction factor for effect of pressure on steel. See API Manual of Petroleum Measurement Standards, Chapter 12, Section 2

C_g = gravity correction factor for orifice well tester to change from a gas relative density of 0.6

C_{tl} = liquid temperature correction factor. Proportional to the thermal coefficient which varies with density and temperature

C_{ts} = correction factor for effect of temperature on steel

C_u = velocity of sound in the gas non-flowing condition.

d = orifice diameter, in.

D = internal pipe diameter of orifice meter run or prover section, in.

DL = Minimum downstream meter tube length, in.

D_p = the difference between the flowing pressure and the equilibrium vapor pressure of the liquid.

D_u = diameter of the meter bore.

ρ = flowing fluid density, lb/ft^3

e = orifice edge thickness, in.

E = orifice plate thickness, in.

E_m = modulus of elasticity for steel $[(30)(10^6)]$ psi

F = liquid compressibility factor

F_a = orifice thermal expansion factor. Corrects for the metallic expansion or contraction of the orifice plate. Generally ignored between 0° and 120°F

F_c = orifice calculation factor

F_g = relative density (specific gravity) factor applied to change from a relative density of 1.0 (air) to the relative density of the flowing gas

F_{gt} = gravity-temperature factor for liquids

F_n = numeric conversion factor

F_{na} = units conversion factor for pitot tubes

F_{pb} = pressure base factor applied to change the base pressure from 14.73 psia

F_{pm} = pressure factor applied to meter volumes to correct to standard pressure. See API Manual of Petroleum Measurement Standards, Chapter 12, Section 2

F_{pv} = supercompressibility factor required to correct for deviation from the ideal gas laws = $\sqrt{1/Z}$

F_s = steam factor

F_{sl} = orifice slope factor

F_{tb} = temperature base factor. To change the temperature base from 60°F to another desired base

F_{tf} = flowing temperature factor to change from the assumed flowing temperature of 60°F to the actual flowing temperature

F_{tm} = temperature correction factor applied to displacement meter volumes to correct to standard temperature. See API Manual of Petroleum Measurement Standards, Chapter 12, Section 2

G, G_l = relative density (specific gravity) at 60°F

G_f = relative density (specific gravity) at flowing temperature

H = pressure, inches of mercury

h_m = differential pressure measured across the orifice plate in inches of mercury at 60°F

h_w = differential pressure measured across the orifice plate in inches of water at 60°F

$\sqrt{h_w P_f}$ = pressure extension. The square root of the differential pressure times the square root of the absolute static pressure

IM = indicated mass, mass measured at flowing conditions without correction for meter performance, counts / K. Also the difference between opening and closing meter readings.

IV = indicated volume, volume measured at flowing conditions without correction for meter performance, counts / K. Also the difference between opening and closing meter readings.

k = ratio of the specific heat at constant pressure to the specific heat at constant volume

K = a numerical constant. Pulses generated per unit volume or mass through a turbine, positive displacement, Coriolis or ultrasonic meter

Fig. 5.2a—General nomenclature of terms found in Chapter 5 (GPSA 2012).

Nomenclature (Cont'd)

Key	$= F_n (F_c + F_{sl}) =$ orifice factor
L	= distance between upstream and downstream transducer.
LTB	= Length of tube bundle, in flow conditioner, in. (See Fig. 3-3)
MF	= meter factor, a number obtained by dividing the actual quantity of liquid passed through the meter during proving by the quantity indicated or registered by the meter
N	= number of whole meter pulses per single proving run
N_{avg}	= average number of pulses or interpolated pulses for proving runs that satisfy the repeatability requirements
P	= pressure, psia
P_b	= base pressure, psia
Pf_1, Pf_2	= static pressure at either the upstream(1) or downstream(2) pressure tap, psia
P_p	= operating or observed pressure, psig.
P_s	= pressure at which the base volume of a meter prover was determined, usually 0 psig.
P_1/P_2	= pressure ratio
Q	= gas flow rate, cu ft/day
Q_h	= rate of flow, std. cu ft/hr or gal./hr
R_b	= maximum differential range, in. of water
R_p	= maximum pressure range of pressure spring, psi
S	= square of supercompressibility

SEP	= Separation distance between elbows, in. (See Fig. 3-9)
SPU	= Signal Processing Unit, the electronic microprocessor system of the multi-path ultrasonic meter.
t1	= transit time from upstream transducer to downstream transducer.
t2	= transit time from downstream transducer to upstream transducer.
T_b	= base temperature, °F
T_f	= flowing temperature, °F
T_{meas}	= reference temperature of the orifice plate bore, °F
UL	= Minimum upstream meter tube length, in. (See Fig. 3-9)
UL1	= See Fig. 3-9
UL2	= See Fig. 3-9
UM	= ultrasonic meter for measuring gas flow rates
V	= velocity of flowing gas.
VOS	= velocity of sound
X	= axial distance separating transducers
Y	= expansion factor to compensate for the change in density as the fluid passes through an orifice
Y_{CR}	= critical flow constant
Z	= compressibility factor
β	= ratio of the orifice or throat diameter to the internal diameter of the meter run, dimensionless
ρ	= density, lb/cu ft or lb/gal

Fig. 5.2b—General nomenclature of terms found in Chapter 5, cont'd (GPSA 2012).

Standard Orifice Designation	Orifice Area, in.²	Valve Body Size (Inlet Diameter x Outlet Diameter), in.										
		1 x 2	1.5 x 2	1.5 x 2.5	1.5 x 3	2 x 3	2.5 x 4	3 x 4	4 x 6	6 x 8	6 x 10	8 x 10
D	0.110	•	•	•								
E	0.196	•	•	•								
F	0.307	•	•	•								
G	0.503			•	•	•						
H	0.785				•	•						
J	1.287					•	•	•				
K	1.838							•				
L	2.853							•	•			
M	3.60								•			
N	4.34								•			
P	6.38								•			
Q	11.05									•		
R	16.0									•	•	
T	26.0											•

Table 5.6—Pressure-relief-valve orifice designations (PetroWiki 2015).

Turbine Meter Equation for Liquid

$$q_b = q_f M_f F_t F_p$$

q_b	Flow rate at base conditions
q_f	Flow rate at operating conditions (meter reading)
M_f	Meter factor to correct meter output based on calibration
F_t	Factor to control fluid, from flowing temperature to base temperature
F_p	Factor to correct fluid, from flowing pressure to base pressure

5.7 Storage Facilities

Minimum Thickness of Shell Plates

$$t_d = \frac{2.6D(H-1)G}{S_d} + C_A$$

$$t_t = \frac{2.6D(H-1)}{S_t}$$

Bottom-Course Thickness

$$t_{1d} = \left(1.06 - \frac{0.463D}{H}\sqrt{\frac{HG}{S_d}}\right)\left(\frac{2.6DHG}{S_d}\right) + C_A$$

$$t_{1t} = \left(1.06 - \frac{0.463D}{H}\sqrt{\frac{H}{S_t}}\right)\left(\frac{2.6DH}{S_t}\right)$$

Upper-Course Thickness

Lowest Value Obtained From the Following:

$$x_1 = 0.61(rt_u)^{0.5} + 3.84CH$$

$$x_2 = 12CH$$

$$x_3 = 1.22(rt_u)^{0.5}$$

Minimum Thickness for Upper-Shell Courses

$$t_{dx} = \frac{2.6D\left(H - \frac{x}{1,000}\right)G}{S_d} + C_A$$

$$t_{tx} = \frac{2.6D\left(H - \frac{x}{1,000}\right)}{S_t}$$

t_d Design shell thickness, in.
t_t Hydrostatic test shell thickness, in.
t_{1d} Design bottom-course thickness, in.
t_{1t} Hydrostatic bottom-course thickness, in.
t_{dx} Design upper-shell course thickness, in.
t_{tx} Hydrostatic upper-shell course thickness, in.
D Nominal tank diameter, ft
H Design liquid level
x Minimum value point for which the calculation is run
G Design-specific gravity of the liquid to be stored, as specified by the purchaser
C_A Corrosion allowance, in., as specified by the purchaser
S_d Allowable stress for the design condition, lbf/in.²
S_t Allowable stress for the hydrostatic test condition, lbf/in.²
t_u Thickness of the upper course at the girth joint, exclusive of any corrosion allowance, in.
C $= \left[K^{0.5}(K-1)\right]/\left(1+K^{1.5}\right)$
K $= t_L / t_u$
t_L Thickness of the lower course at the girth joint, exclusive of any corrosion allowance, in.
r Nominal tank radius, in.

Minimum Width of the Tank-Bottom Reinforcing Plate

$$t_b = \frac{h^2}{14,000} + \frac{b}{310}\sqrt{HG}$$

t_b Minimum thickness of the bottom reinforcing plate, in.
h Vertical height of clear opening, in.
b Horizontal width of clear opening, in.
H Maximum design-liquid level, ft
G Specific gravity, not less than 1.0

Minimum Section Modulus of the Stiffening Ring

$$Z = 0.0001 D^2 H_2 \left(\frac{V}{120}\right)^2$$

Z Required minimum section modulus, in.^{3}
D Nominal tank diameter, ft
H_2 Height of the tank shell, ft
V Design wind speed (3-sec gust), mph

Maximum Height of the Unstiffened Shell

$$H_1 = 600,000 \, t \sqrt{\left(\frac{t}{D}\right)^3 \left(\frac{120}{V}\right)^2}$$

H_1 Vertical distance, ft,
t As-built thickness of the thinnest shell course, in.
D Nominal tank diameter, ft
V Design wind speed (3-sec gust), mph

Allowable Stress on Roof-Supporting Columns

$$F_a = \frac{\left\{\dfrac{\left[1 - \dfrac{(l/r)^2}{2C_c^2}\right]}{\dfrac{5}{3} + \dfrac{3(l/r)}{8C_c} - \dfrac{(l/r)^3}{8C_c^3}}\right\}}{\left(1.6 - \dfrac{l}{200r}\right)} \qquad l/r \leq C_c$$

$$F_a = \frac{\left[\dfrac{12\pi^2 E}{23(l/r)^2}\right]}{\left(1.6 - \dfrac{l}{200r}\right)} \qquad l/r > C_c$$

$$C_c = \sqrt{\frac{2\pi^2 E}{F_y}}$$

F_a Allowable compression stress, lbf/in.^{2}
F_y Yield stress of material, lbf/in.^{2}
E Modulus of elasticity, lbf/in.^{2}
l Unbraced length of the column, in.
r Least radius of gyration of column, in.
C_c Theoretical demarcation line between inelastic and elastic behavior

Rafters Requirement

$$b = t\left(1.5F_y / p\right)^{\frac{1}{2}} \le 84 \text{ in.}$$

b Maximum allowable roof plate span, in.
F_y Specified minimum-yield strength of roof plate
t Corroded roof thickness, in.
p Uniform pressure, psia

Self-Supporting Cone Roofs

$$\text{minimum thickness} = \text{greatest of } \frac{D}{400 \sin\theta}\sqrt{\frac{T}{45}} + CA, \frac{D}{460\sin\theta}\sqrt{\frac{U}{45}} + CA, \text{ and } \tfrac{3}{16} \text{ in.}$$

maximum thickness = $\frac{1}{2}$ in., exclusive of corrosion allowance

D Nominal diameter of the tank shell, ft
T Greater of load combinations with balanced snow load, S_b, lbf/ft^2
U Greater of load combinations with unbalanced snow load, S_u, lbf/ft^2
θ Angle of cone elements to the horizontal line, deg
CA Corrosion allowance

Self-Supporting Dome and Umbrella Roofs

$$\text{minimum thickness} = \text{greatest of } \frac{r_r}{200}\sqrt{\frac{T}{45}} + CA, \frac{r_r}{230}\sqrt{\frac{U}{45}} + CA, \text{ and } \tfrac{3}{16} \text{ in.}$$

maximum thickness = $\frac{1}{2}$ in., exclusive of corrosion allowance

T Greater of load combinations with balanced snow load, S_b, lbf/ft^2
U Greater of load combinations with unbalanced snow load, S_u, lbf/ft^2
r_r Roof radius, ft

Liquid Weight at the Shell

$$w_L = 4.67 t_b \sqrt{F_{by} / H}$$

w_L Liquid weight, lbf/ft
F_{by} Minimum specified-yield stress of the bottom plate under the shell, lbf/in.^{2}
H Design liquid height, ft
t_b Required thickness of the bottom plate under the shell, in.

Net Uplift Loads

Design Pressure

$$\left[\left(P - 8t_h\right) \times D^2 \times 4.08\right] - W_1$$

Test Pressure

$$\left[\left(P_t - 8t_h\right) \times D^2 \times 4.08\right] - W_1$$

Failure Pressure

$$\left[\left(1.5P_f - 8t_h\right) \times D^2 \times 4.08\right] - W_3$$

Wind Load

$$P_{WR} \times D^2 \times 4.08 + \left(4M_w / D\right) - W_2$$

Seismic Load

$$\left(4\times M_s\,/\,D\right)-W_2\left(1-0.4A_V\right)$$

Design Pressure + Wind

$$\left[\left(0.4P+P_{WR}-0.08t_h\right)\times D^2\times4.08\right]+\left(4M_{WH}\,/\,D\right)-W_1$$

Design Pressure + Seismic

$$\left[\left(0.4P-0.08t_h\right)\times D^2\times4.08\right]+\left(4M_s\,/\,D\right)-W_1\left(1-0.4A_V\right)$$

Frangibility Pressure

$$\left[\left(3P_f-8t_h\right)\times D^2\right]-W_3$$

A_V Vertical earthquake acceleration coefficient, % g
D Tank diameter, ft
H Tank height, ft
M_{WH} $=P_{WS}\times D\times H^2\,/\,2$, ft-lbf
M_s Seismic moment, ft-lbf
P Design pressure, in. of water column, psia
P_f Failure pressure, in. of water column, psia
P_t Test pressure, in. of water column, psia
P_{WR} Wind uplift pressure on roof, in. of water column, psia
P_{WS} Wind pressure on shell, lbf/ft^2
t_h Roof plate thickness, in.
W_1 Dead load of shell minus any corrosion allowance, and any dead load other than roof plate acting on the shell minus any corrosion allowance, lbf
W_2 Dead load of shell minus any corrosion allowance, and any dead load including roof plate acting on the shell minus any corrosion allowance, lbf
W_3 Dead load of the shell using as-built thicknesses, and any dead load other than roof plate acting on the shell using as-built thickness, lbf

Shell and Tube Heat Transfer

Basic Convention Heat-Transfer Equation

$$Q=U\ A\ \Delta T_m$$

Q Heat transferred, Btu/hr
U Overall heat-transfer rate, Btu/ hr-ft^2,°F
A Outside heat-transfer surface of tubes, ft^2
ΔT_m Corrected mean temperature difference, °F

Log Mean Temperature Difference

$$LMTD=\frac{\left(T_{hot,\,out}-T_{cold,\,in}\right)-\left(T_{hot,\,in}-T_{cold,\,out}\right)}{\ln\left(\dfrac{T_{hot,\,out}-T_{cold,\,in}}{T_{hot,\,in}-T_{cold,\,out}}\right)}$$

$LMTD$ Log mean temperature difference
T Temperature, °F

Overall Heat-Transfer Rate (U)

$$U=\frac{1}{\Sigma r}$$

$$\sum r=r_{ft}+r_{fs}+r_m+r_{dt}+r_{ds}$$

$$r_m = \frac{t_w D_o}{12 k_m \left(D_o - 2t_w \right)}$$

r	Resistance
r_{ft}, r_{fs}	Fouling-resistance tube and shell side, respectively
r_{dt}, r_{ds}	Film-resistance tube and shell side, respectively
r_m	Metal resistance
t_w	Tube wall thickness, in.
k_m	Tube metal thermal conductivity, Btu/hr-ft^2 (°F/ ft)
D_o	Metal resistance

Storage Pressure

$$\phi = P + \left[(\Delta - p) \frac{(T + 460)}{(t + 460)} \right] - A$$

ϕ	Required storage pressure, psig
P	Vapor pressure of liquid at maximum surface temperature, psia
p	Vapor pressure of liquid at minimum surface temperature, psia
Δ	Absolute pressure in tank at which vacuum vent opens, psia
T	Maximum average temperature of air-vapor mixture, °F
t	Minimum average temperature of air-vapor mixture, °F
A	Atmospheric pressure = 14.7 psia

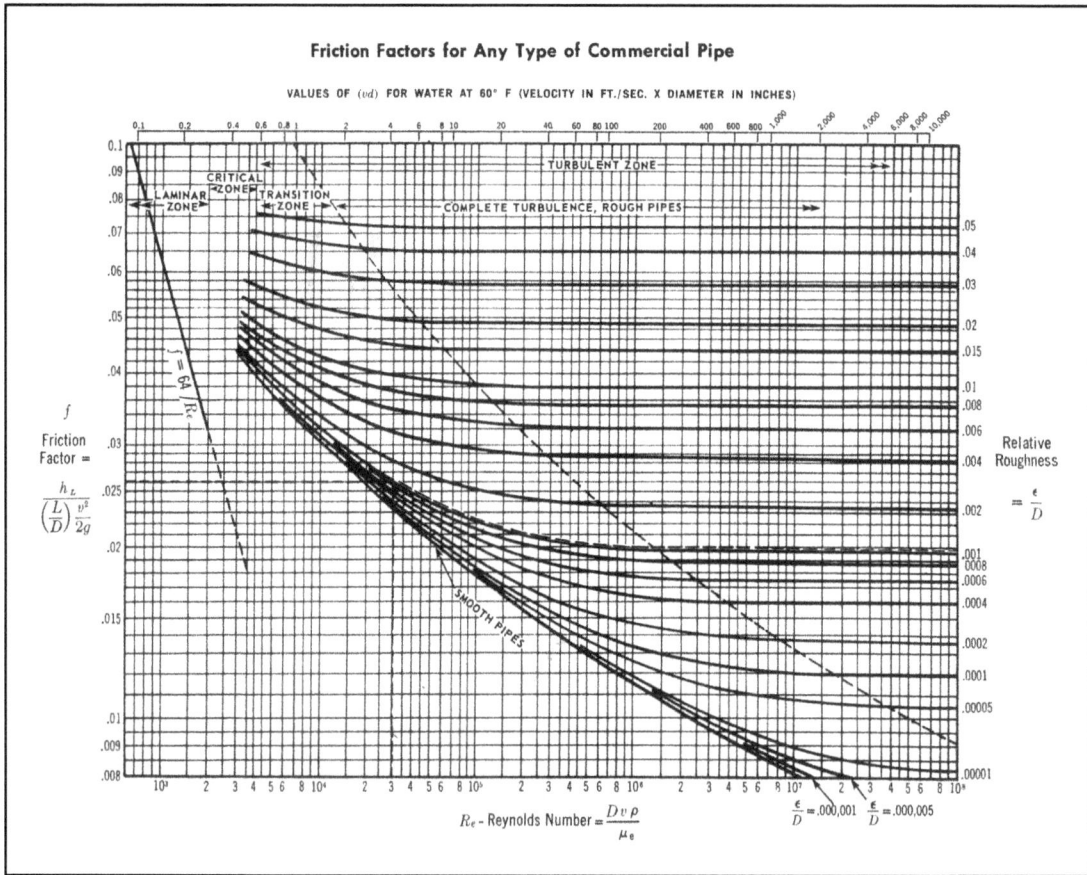

Fig. 5.3—Friction factors for commercial pipe (Crane 1972).

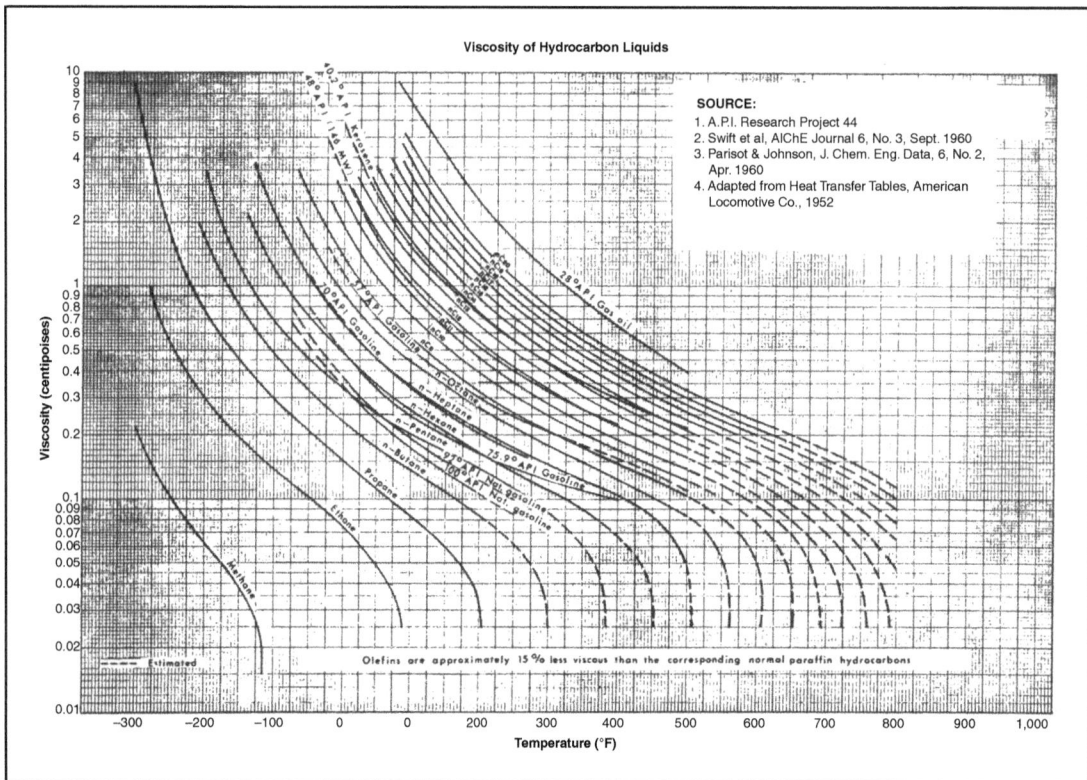

Fig. 5.4—Viscosities of hydrocarbon liquids (GPSA 2012).

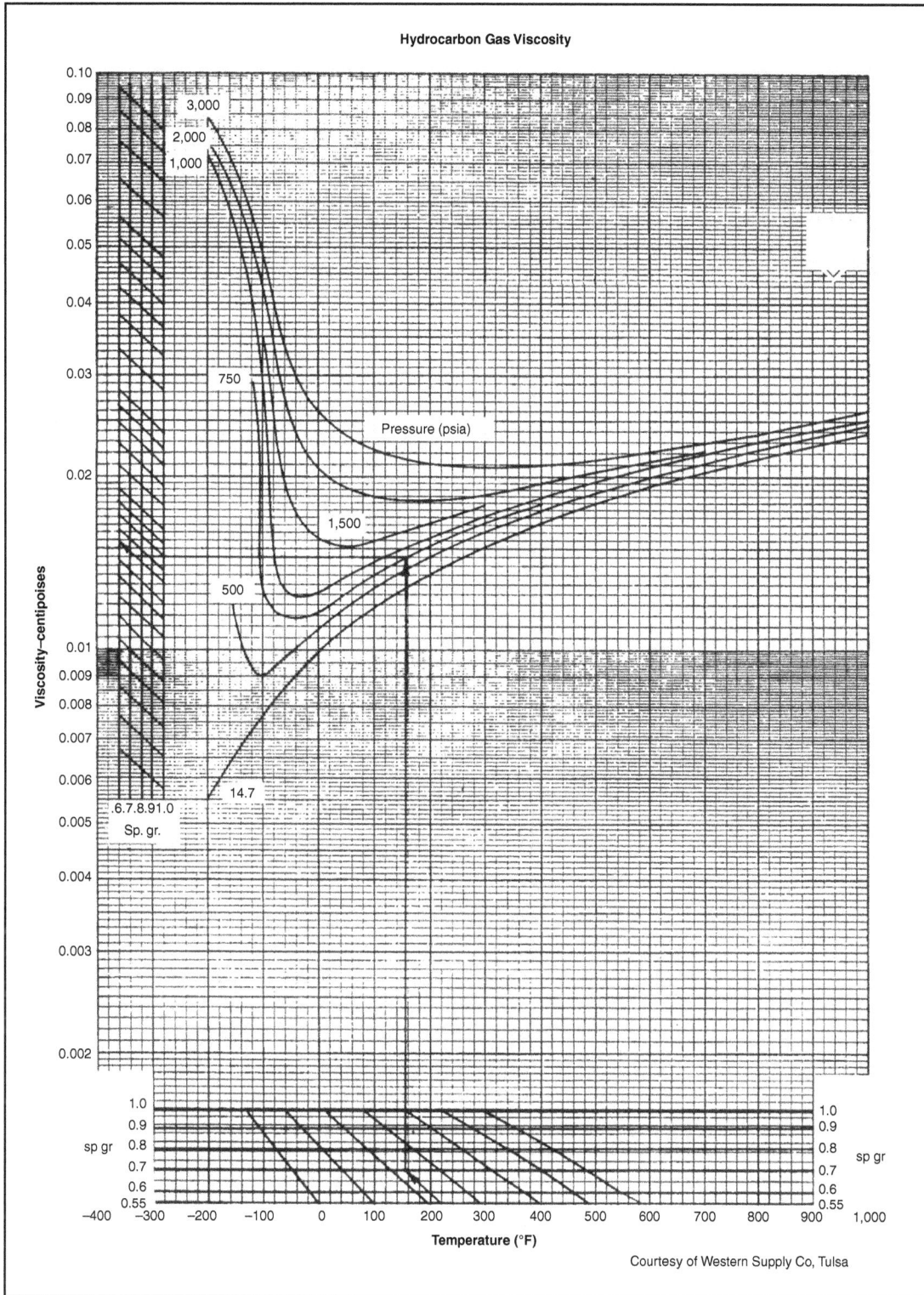

Fig. 5.5—Hydrocarbon-gas viscosity vs. temperature (GPSA 2012).

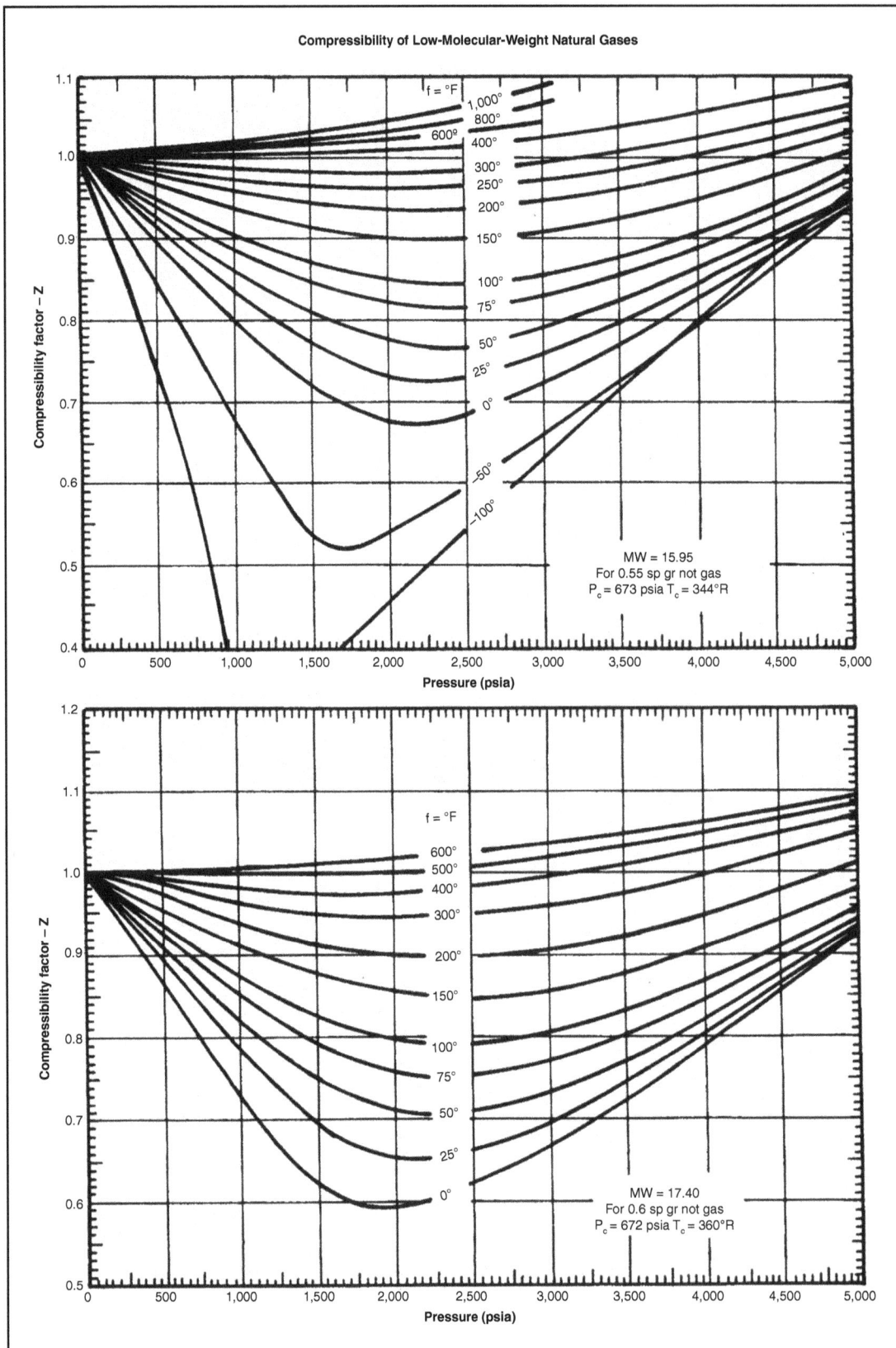

Fig. 5.6a—Compressibility of low-molecular-weight natural gases (GPSA 2012).

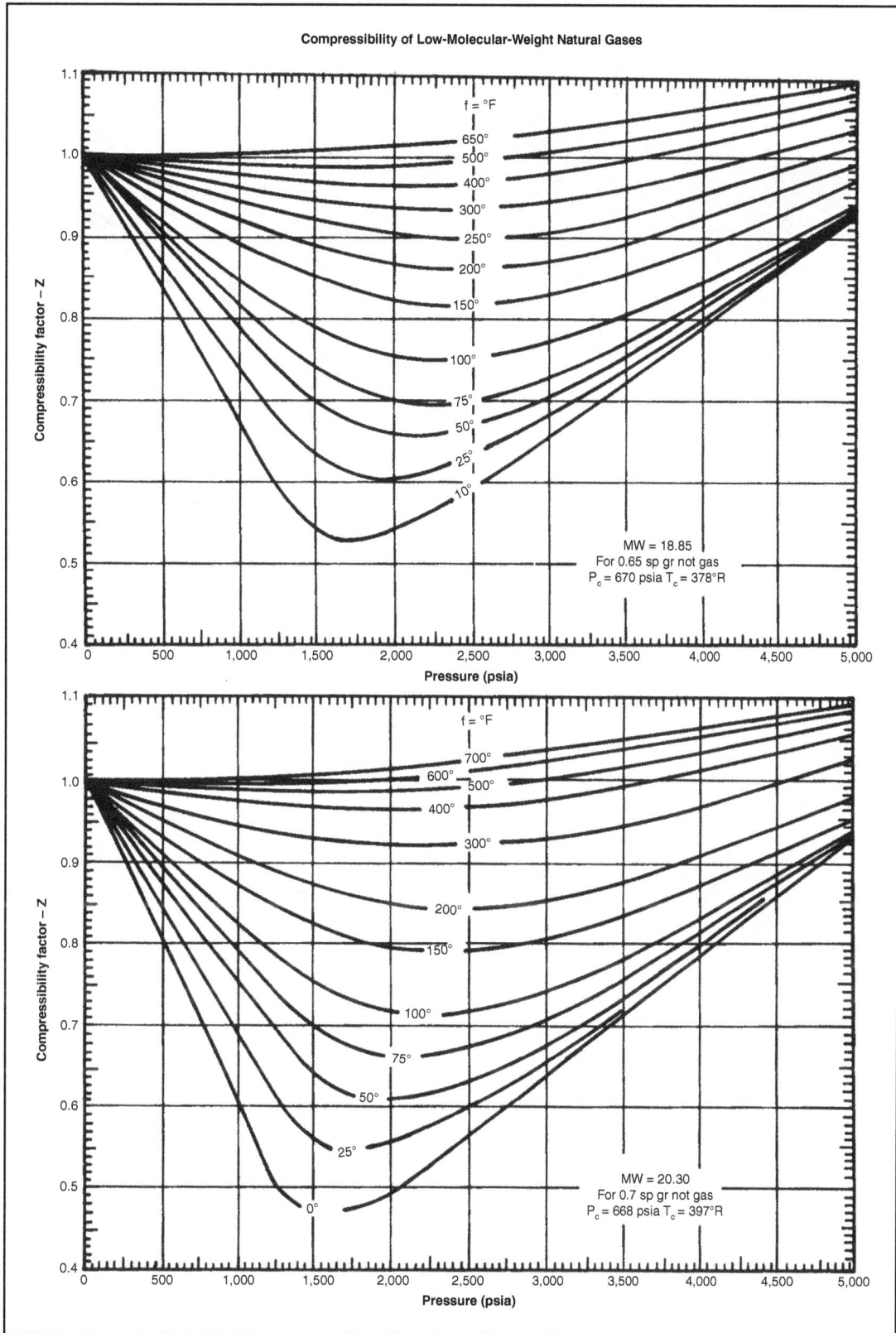

Compressibility of Low-Molecular-Weight Natural Gases

Fig. 5.6b—Compressibility of low-molecular-weight natural gases, cont. (GPSA 2012).

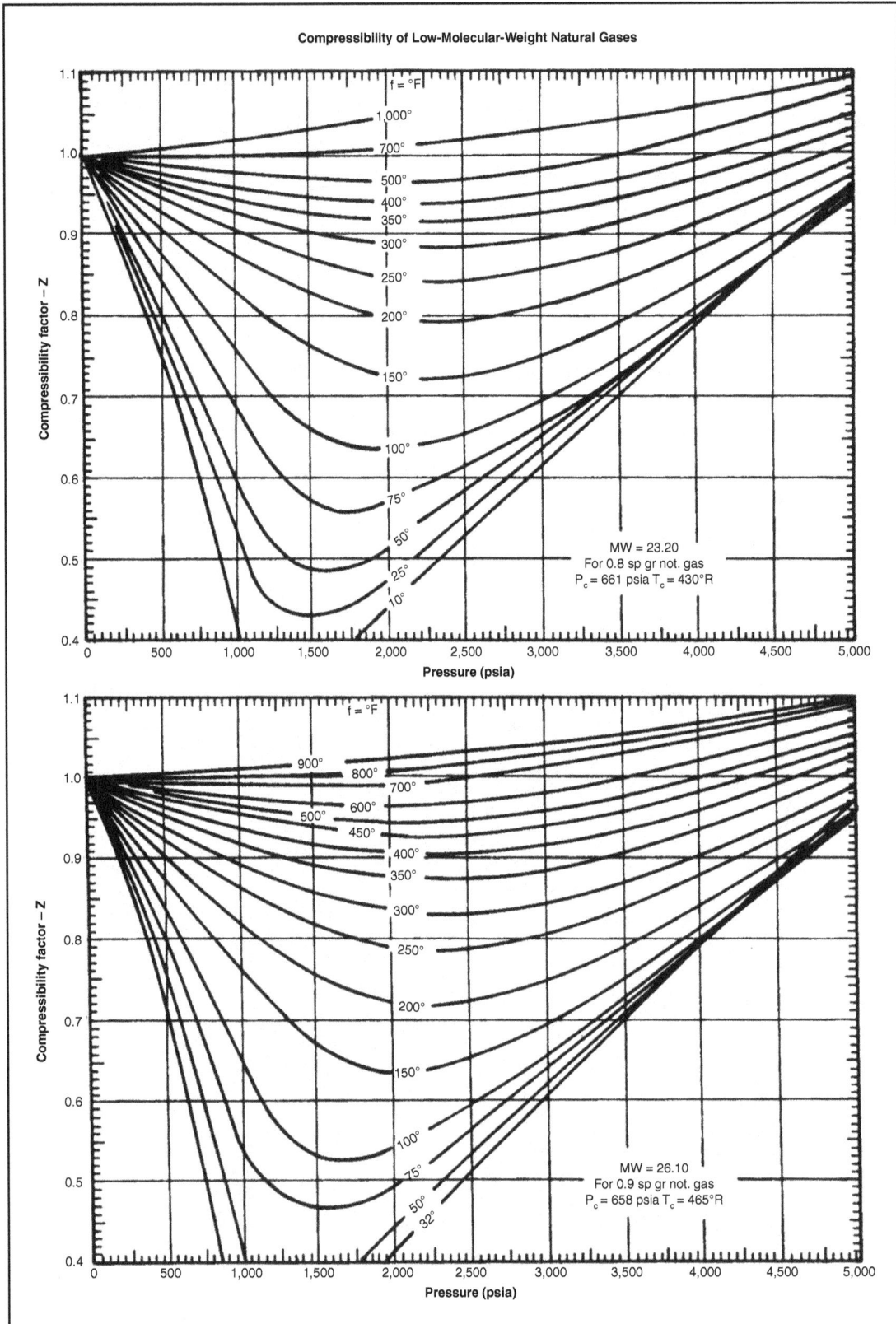

Fig. 5.6c—Compressibility of low-molecular-weight natural gases, cont. (GPSA 2012).

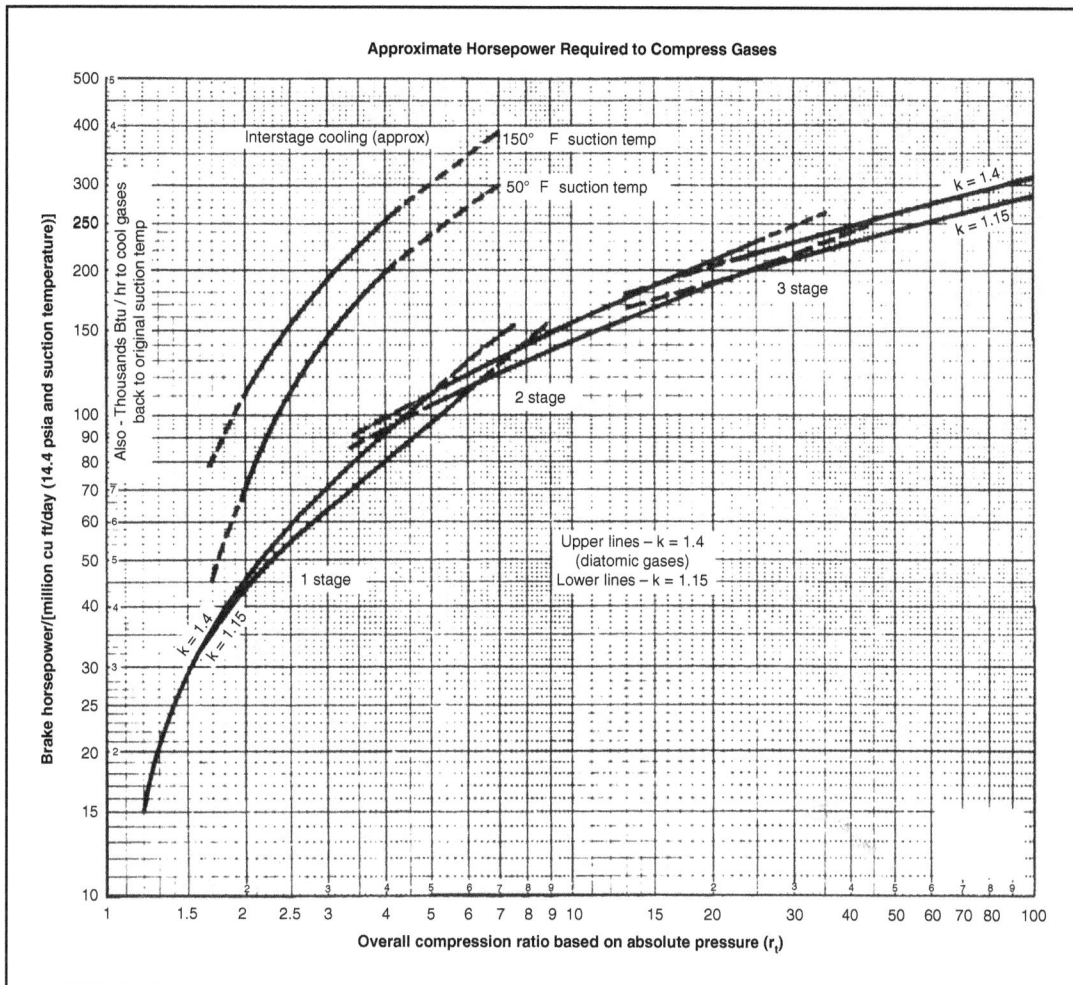

Fig. 5.7—Approximate horsepower required to compress gases (GPSA 2012).

185

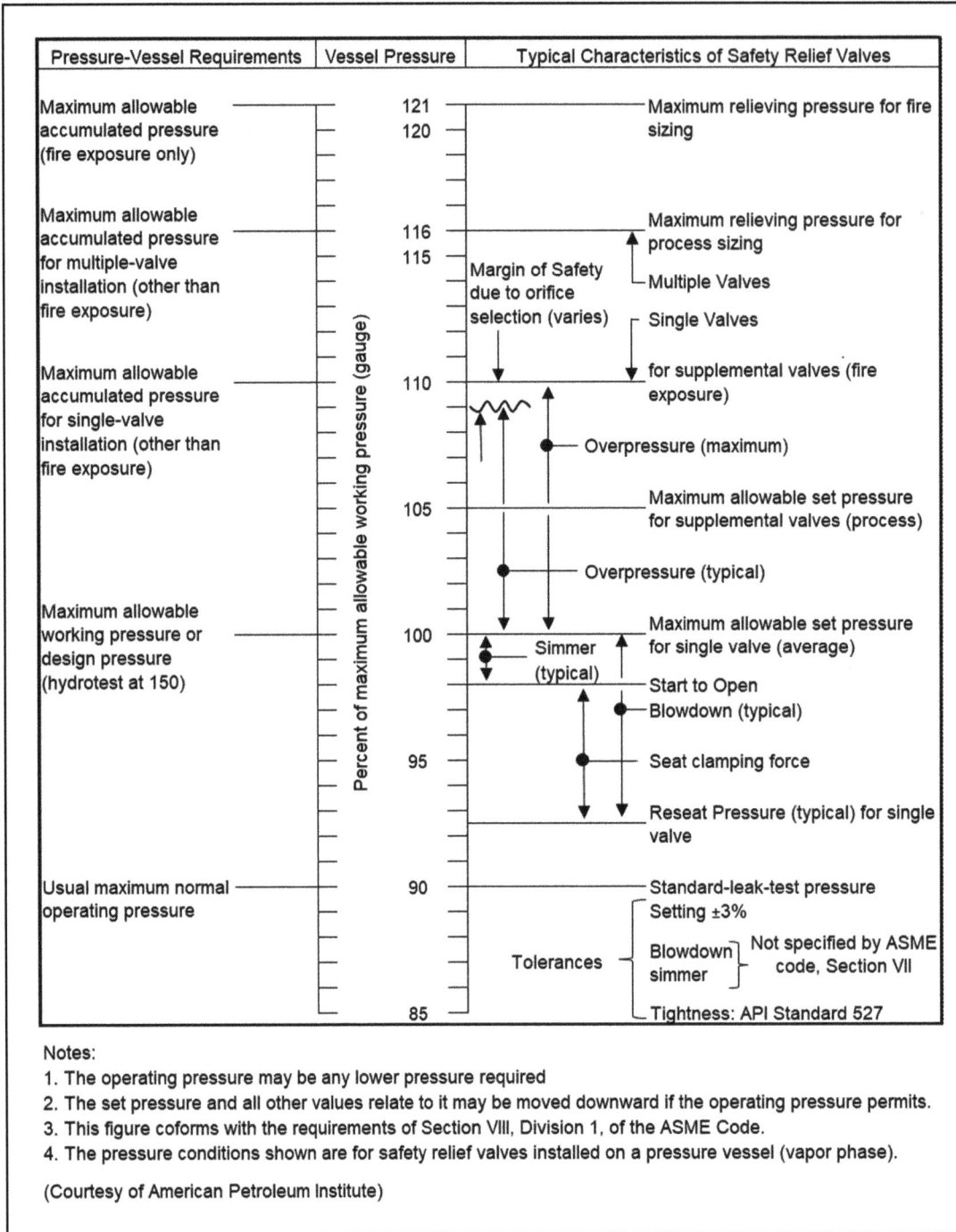

Pressure-Vessel Requirements	Vessel Pressure	Typical Characteristics of Safety Relief Valves
Maximum allowable accumulated pressure (fire exposure only)	121 / 120	Maximum relieving pressure for fire sizing
Maximum allowable accumulated pressure for multiple-valve installation (other than fire exposure)	116 / 115	Maximum relieving pressure for process sizing — Multiple Valves — Single Valves
Maximum allowable accumulated pressure for single-valve installation (other than fire exposure)	110	for supplemental valves (fire exposure) — Overpressure (maximum)
	105	Maximum allowable set pressure for supplemental valves (process)
		Overpressure (typical)
Maximum allowable working pressure or design pressure (hydrotest at 150)	100	Maximum allowable set pressure for single valve (average) — Simmer (typical) — Start to Open — Blowdown (typical)
	95	Seat clamping force — Reseat Pressure (typical) for single valve
Usual maximum normal operating pressure	90	Standard-leak-test pressure
		Tolerances — Setting ±3% — Blowdown simmer — Not specified by ASME code, Section VII
	85	Tightness: API Standard 527

Margin of Safety due to orifice selection (varies)

Percent of maximum allowable working pressure (gauge)

Notes:

1. The operating pressure may be any lower pressure required
2. The set pressure and all other values relate to it may be moved downward if the operating pressure permits.
3. This figure coforms with the requirements of Section VIII, Division 1, of the ASME Code.
4. The pressure conditions shown are for safety relief valves installed on a pressure vessel (vapor phase).

(Courtesy of American Petroleum Institute)

Fig. 5.8—Multiple pressure-relief-valve requirements (courtesy of API).

Chapter 6

Petroleum Economics

6.1 Valuation of Oil and Gas Reserves

Annual Discounting

$$D_f = \frac{1}{\left(1+r\right)^t}$$

Monthly Discounting

$$D_f = \frac{1}{\left(1+\dfrac{r}{12}\right)^{12t}}$$

Continuous Discounting

$$D_f = \frac{1}{e^{rt}}$$

D_f Discount factor for a specific point in time
r Annual rate of discount expressed as a fraction or percent
t Time in years from the as-of date to the point of reference for discounting

Weighted Average Cost of Capital

$$C_{at} = C_{bt} \times \left(1 - T_R\right)$$

C_{at} Weighted average cost of capital after federal income tax
C_{bt} Weighted average cost of capital before federal income tax
T_R Tax rate

Future Value of Lump Sum

$$F = P(1+i)^n$$

Present Value of Lump Sum

$$P = \frac{F}{(1+i)^n}$$

Sinking Fund

$$A = F\left[\frac{i}{(1+i)^n - 1}\right]$$

Annuity (Loan)

$$A = P\left[\frac{i}{(1+i)^n - 1} + 1\right]$$

Annuity Due or Lease

$$B = P\left[\frac{i}{(1+i)^n - 1} + (1+i)^{n-1}\right]$$

Modified Sinking Fund

$$B = F\left[\frac{i}{(1+i)^n - 1} / (1+i)\right]$$

Nominal Interest Rates

$$i_n = i \times m$$

Effective Interest Rates

$$i_e = (1+i)^m - 1$$

i	Interest rate per period
n	Total number of interest periods
m	Total number of compounding periods
P	Present lump sum of money
F	Future lump sum of money (received n periods from now)
A	End-of-period payment or receipt in a uniform series continuing for n periods
B	Beginning-of-period payment or receipt in a uniform series continuing for n periods

6.2 Reservoir Management

Net Cash Flow (NCF) = Net Annual Revenue – Net Annual Expenditure = Total Cash In – Total Cash Out

Capital Expenditures (CAPEX) = Beginning Net Fixed Assets – Ending Net Fixed Assets – Depreciation Expense

Operating Expenditures (OPEX) = Production Cost per Unit, USD/bbl × Production Volume, bbl

Or

Lifting Cost per Unit, USD/bbl = (OPEX + Royalties + Exploration Expenses + Depreciation)/Sales Volume

Net Profit = Revenues − Costs = Cash Receipts − Cash Disbursement

Discount Cash Flow = Net Cash Flow × Discount Factor

Discount Cash Out (CAPEX + OPEX) = Cash Out × Discount Factor

Discounted Production, bbl = Daily Production × (365/1,000,000) × Discount Factor

Dividend Payout Ratio = Dividend per Share / Earnings per Share

Return On Investment (ROI) = Σ NCF / Investment

Return On Assets (ROA) = Average Net Income / Average Book Investment

Discounted Rate On Investment (DROI) = NPV / Present Value of Investment

Weighted Average Cost of Capital

$$C_{RW} = \frac{C_{RD}\left(1 - T_R\right)C_D + C_{RE}C_E + C_{RP}C_P}{C_D + C_E + C_P}$$

C_{RW} Weighted average cost of capital, %
C_{RD} Cost of debt, %
C_{RE} Cost of common stock equity, %
C_{RP} Cost of preferred stock, %
C_D Market value of debt
C_E Market value of common stock equity
C_P Market value of preferred stock
T_R Tax rate

Depreciation, Depletion, and Amortization Formulas

Straight Line Depreciation (C_{SL})

$$C_{SL} = C / t_L$$

C Cost of tangible capital item
t_L Depreciation life

Sum-of-the-Years Depreciation (SYD_n)

$$SYD_n = \left[\frac{t_L - n + 1}{\left(\sum_{n=1}^{t_L} n\right)}\right] \times C$$

C Cost of tangible capital item
t_L Depreciation life
n Depreciation year

Hyphenate: Declining-Balance Depreciation (DB_i)

$$\sum_{i=1}^{n} DB_i = C\left\{1 - \left[1 - \left(M / t_L\right)\right]^n\right\}$$

C Cost of tangible capital item
t_L Depreciation life
M Multiplier (for double-declining balance = 2)
n Number of years since start of depreciation

189

Switch From Declining Balance to Straight Line

$$N = t_L - \left(t_L / M \right) + 1$$

N	Year to switch from declining balance to straight line
t_L	Depreciation life
M	Multiplier (for double-declining balance = 2)

Unit-of-Production (UOP) Depreciation or Cost Depletion

$$UOP_n = C\left(Q_n / U \right) \qquad or \qquad UOP_n = \left(C - \sum_{i=1}^{n} UOP_i \right) \times \left(Q_n / R_n \right)$$

C	Cost of tangible capital item or leasehold cost
Q_n	Annual production (BOE) of year n
U	Ultimate recovery, $= \sum_{n=1}^{t_L} = Q_n$
R_n	Reserves at beginning of year n
n	Depreciation year
t_L	Producing life

6.3 Investment Decision Analysis

Accounting Rate of Return

$$ARR = \frac{NI}{(I - S)/2}$$

$$ARR = \frac{NI}{I}$$

ARR	Accounting rate of return
NI	Average financial income (after taxes) over the estimated life of the project
I	Initial capital investment
S	Salvage value

Net Present Value

$$NPV = \sum_{j=0}^{L} \frac{NCF_j}{\left(1 + i \right)^j}$$

NPV	Net present value discounted at rate i
NCF_j	Net cash flow for period j
L	Project life
i	Rate of return

Discounted Profit-to-Investment Ratio

$$\frac{P}{I} = \frac{\sum_{j=0}^{L} \frac{NCF_j}{\left(1 + i_{op} \right)^j}}{\sum_{j=0}^{L} \frac{INV_j}{\left(1 + i_{op} \right)^j}}$$

$$\frac{P}{I} = \frac{\sum_{j=1}^{L} \dfrac{NOI_j}{\left(1+i_{op}\right)^j}}{\sum_{j=0}^{L} \dfrac{INV_j}{\left(1+i_{op}\right)^j}}$$

P   Profit
I   Investment

Discounted Cash Flow Rate of Return

$$\sum_{j=0}^{L} \frac{INV_j}{\left(1+i\right)^j} = \sum_{j=1}^{L} \frac{NOI_j}{\left(1+i\right)^j}$$

NOI_j Net operating income period j
NCF_j Net cash flow period j
INV_j Investment period j
i_{op} Average investment opportunity rate
i Discounted cash flow rate of return
L Life of the project

Appreciation of Equity Rate of Return

$$i = \left[\frac{\sum_{j=1}^{L} NOI_j \left(1+i_{op}\right)^{L-j}}{\sum_{j=0}^{L} \dfrac{INV_j}{\left(1+i_{op}\right)^j}} \right]^{\frac{1}{L}} - 1$$

NOI_j Net operating income period j
INV_j Investment period j
i_{op} Average investment opportunity rate
i Appreciation of equity rate of return
L Terminal point

Liquified Natural Gas Screening

$$n = \frac{L}{5,000} + .25$$

n Number of ships required for 1 million tons per annum (mtpa)
L One-way distance in nautical miles

6.4 Risk Analysis

Measures of Risk

Risk-adjusted capital: Maximum amount of money that can be lost (with a certain confidence)

Value at risk: Difference between the mean and the maximum amount of money that can be lost (with a certain confidence)

Return on risk-adjusted capital: Relation between expected profit and the maximum amount of money that can be lost

Expected value (EV) = Probability-weighted value of all possible outcomes

Expected monetary value (EMV) = Expected value of the present values of the net cash flows = EV (NPV)

Or

$$EMV = \sum_{all\ i} P\left(\text{outcome } i\right) \times NPV_{\text{outcome } i}$$

$$\text{Risked } DROI = \frac{EMV}{EV\left(PV \text{ of Investment}\right)}$$

Judging Probability (P) of Recovery

P (wildcat discovery) = P(trap) × P(source) × P(porosity and permeability) × etc.

Profit or Net Cash Flow Calculations for a Royalty-Tax System

Gross production – shrinkage =

Gross sales × net revenue interest =

Net sales × price =

Gross revenue – share of state and local taxes – share of operating costs =

Net operating income before income taxes – income tax =

Net operating income after income taxes – share of investments =

Net cash flow after income tax

Measures of Dispersion

Population Variance

$$= \frac{1}{N} \sum \left(x_i - \overline{x}\right)^2$$

Sample Variance

$$= \frac{1}{N-1} \sum \left(x_i - \overline{x}\right)^2$$

Skewness

$$= \left[\frac{N}{\left(N-1\right)\times\left(N-2\right)}\right]\left(\frac{1}{s^3}\right) \sum \left(x_i - \overline{x}\right)^3$$

Normal Distribution

$$f\left(x\right) = \frac{1}{\sqrt{2\pi\sigma^2}} \exp\left[-\frac{\left(x-\mu\right)}{2\sigma^2}\right]$$

Log-Normal Distribution

$$f\left(x\right) = \frac{1}{\sqrt{2\pi\sigma_1^{\ 2}}} \exp\left[-\frac{\left(\ln x - \mu_1\right)}{2\sigma_1^{\ 2}}\right] \qquad \mu_1 = \ln\left(\frac{\mu^2}{\sqrt{\sigma^2 + \mu^2}}\right) \qquad \sigma_1 = \sqrt{\ln\left(\frac{\sqrt{\sigma^2 + \mu^2}}{\mu^2}\right)}$$

N	Number in population
x	Discrete value in set
μ	Mean
σ	Standard deviation

Swanson's Mean

$$\overline{X}_{\text{Swanson}} = 0.30 P_{90} + 0.40 P_{50} + 0.30 P_{10}$$

$\bar{X}_{Swanson}$ Swanson's mean; approximation of mean for a moderately skewed (usually log-normal) distribution

P_{90} 90th percentile of distribution or 90% chance of exceeding this value, often referred to at P10 low-side value

P_{50} 50th percentile of distribution or 50% chance of exceeding this value, often referred to at P50 medium or median value

P_{10} 10th percentile of distribution or 10% chance of exceeding this value, often referred to at P90 high-side value

Probability Equations

Bayes' Rule

$$P(A_i|B) = \frac{P(B|A_i) \cdot P(A_i)}{\sum_{i=1}^{k} P(B|A_i) \cdot P(A_i)}$$

$P(A_i|B)$ Posterior probabilities
$P(A_i)$ Prior-event probabilities

Binomial Distribution

$$P(x) = C_x^n \cdot p^x \cdot q^{n-x}$$

$$C_x^n = \frac{n!}{x!(n-x)!}$$

$P(x)$ Probability of obtaining exactly x successes in n trials
p Probability of success
q Probability of failure
n Number of trials considered
x Number of successes

Multinomial Distributions

$$P(S) = \frac{N!}{k_1! k_2! \ldots k_m!} P_1^{k_1} P_2^{k_2} \ldots P_m^{k_m}$$

$P(S)$ Probability of the particular sample
P Probabilities of drawing types $1, 2, \ldots m$ from population
N Size of sample $= k_1 + k_2 + \ldots + m$
$k_1, k_2, \ldots k_m$ Total number of outcomes of type $1, 2, \ldots, m$
m Number of different types

Hypergeometric Distributions

$$P(x) = \frac{\binom{C}{x}\binom{N-C}{n-x}}{\binom{N}{n}}$$

$P(x)$ Probability of successes if trials are dependent and selected
N Number of items in the population
C Number of total successes in the population
n Number of trials (size of the sample)
x Number of successes observed in the sample

Poisson Distributions

$$P(x) = \frac{\lambda^x}{x!} e^{-\lambda}$$

$P(x)$ Probability of exactly x occurrences
λ Average number of occurrences per interval of time or space
x Number of occurrences per basic unit of measure

Normal Distributions

$$f(x) = \frac{1}{\sigma\sqrt{2\pi}} e^{-\frac{1}{2}\left(\frac{x-\mu}{\sigma}\right)^2}$$

$f(x)$ Probability density function
μ Mean
σ Standard deviation

Uniform Distributions

$$f(x) = \frac{1}{x_{max} - x_{min}}$$

Triangular Distributions

$$f(x) = \begin{cases} \left(\frac{X - X_{min}}{X_{mod} - X_{min}}\right)^2 \cdot \left(\frac{X_{mod} - X_{min}}{X_{max} - X_{min}}\right), & X_{min} \geq X \geq X_{mod} \\ \left(\frac{X_{max} - X}{X_{max} - X_{mod}}\right)^2 \cdot \left(\frac{X_{max} - X_{mod}}{X_{max} - X_{min}}\right), & X_{mod} \geq X \geq X_{max} \end{cases}$$

Petroleum Exploration Risk Analysis

Probability of Wildcat Discovery =

$(\text{Probability of reservoir trap}) \times (\text{Probability that trap is in the position projected from seismic}) \times$

$(\text{Probability of pay thickness}) \times (\text{Probability of hydrocarbons in formation})$

Probability of x_1, x_2, x_3, ..., x_r outcomes in a sample of n trials = $\dfrac{\left(C_{x_1}^{d_1}\right)\left(C_{x_2}^{d_2}\right)\left(C_{x_3}^{d_3}\right)\cdots\left(C_{x_r}^{d_r}\right)}{\left(C_n^N\right)}$

$$d_1 + d_2 + d_3 + ,\ldots, + d_r = N$$
$$x_1 + x_2 + x_3 + ,\ldots, + x_r = n$$

n Number of trials (wells) in the multiple-well program

N Total number of undrilled prospects in the basin or exploration area before the n wells are drilled

r Number of possible outcomes on any trial

$d_1, d_2, d_3, \ldots, d_r$ Number of elements in the sample space designated as each outcome before the n wells is drilled

$x_1, x_2, x_3, \ldots, x_r$ Number of outcomes that occur in each of the r categories

Probability of x successes in n independent trials $= \left(C_x^n\right)(p)^x (1-p)^{n-x}$

n Number of independent trials

x Number of successes in the n trials, $0 \le x \le n$

p Probability of success on any given trial, $0 \le p \le 1.0$

$\left(C_x^n\right)$ Combination of n things taken x at a time $= \dfrac{n!}{x!(n-x)!}$

Probability of $x_1, x_2, x_3, \ldots, x_r$ outcomes in n independent trials $= \dfrac{n!}{x_1!x_2!x_3!\cdots x_r!(n-x)!}(p_1)^{x_1}(p_2)^{x_2}(p_3)^{x_3}\cdots(p_r)^{x_r}$

$p_1 + p_2 + p_3 + \ldots + p_r = 1.0$

$x_1 + x_2 + x_3 + \ldots + x_r = n$

n Total number of independent trials

r The number of possible outcomes

$x_1, x_2, x_3, \ldots, x_r$ The number of times the outcomes occur in the trials

$p_1, p_2, p_3, \ldots, p_r$ Probability of occurrence of the outcomes on any given independent trial

Risk Analysis Using Simulation Techniques

$$EMV_A = \left[P(\text{Dry})\begin{pmatrix}\text{Dry Hole} \\ \text{Expenses Plus} \\ \text{Write-offs}\end{pmatrix} + P(\text{Gas})\times\begin{pmatrix}\text{Conditional} \\ \text{Mean Value,} \\ \mu_{EMV(\text{Gas})}\end{pmatrix} + P(\text{Oil})\times\begin{pmatrix}\text{Conditional} \\ \text{Mean Value,} \\ \mu_{EMV(\text{Oil})}\end{pmatrix} \right]$$

EMV_A The expected value at chance node A

$P(\text{Dry})$ Probability of dry hole

$P(\text{Gas}), P(\text{Oil})$ Likelihood of occurrence of gas and oil

Chapter 7

Conversion Factors

Length

1 meter = 39.37 in. = 5.0292 rods
1 inch = 2.54 cm
1 feet = 30.48 cm = 0.3048 m = .0606 rods
1 mile = 5,280 ft = 1,760 yards = 1609.344 m
1 nautical mile = 6,076 ft
1 sq mile = 27,878,400 ft^2 = 640 acres = 1 section
1 revolution = 360 degrees = 6.2832 radians = 2π

Mass

1 lbm = 453.6 g = 0.4536 kg = 7,000 gr (grain) = 16 oz
1 kg = 1000 g = 2.2046 lbm
1 slug = 1 lbf·s^2/ft = 32.174 lbm
1 US ton = 2,000 lbm (also called short ton)
1 British ton = 2,240 lbm (also called long ton)
1 tonne = 1,000 kg (also called metric ton) = 2,204.6 lbm
1 kip = 1,000 lbm

Force

1 lbf = 4.448 N = 4.448 × 10^5 dynes
1 lbf = 32.174 poundals = 32.174 lbm·ft/s^2

Gravitational Acceleration

g = 32.2 ft/s^2 = 9.81 m/s^2

$$g_\lambda = 9.7803267714\left(\frac{1+0.00193185138639\ \sin^2\lambda}{\sqrt{1-0.00669437999013\ \sin^2\lambda}}\right)\ \text{m/s}^2$$

λ The geographic latitude of the Earth's ellipsoid measured from the equator, deg

Pressure

1 atm = 14.69595 psia = 2,118 lbf/ft^2
     = 29.92 in. Hg = 760 mm Hg = 1.013 bar
     = 33.93 ft H$_2$O = 1.013 × 10 Pa = 101.3 kPa
1 Pa = 1 N/m^2 = 10^{-5} bar
psia = psig + 14.7

Volume

1 ft^3 = 7.481 US gal = 6.22889 Imperial gal = 28.317 L
1 m^3 = 1,000 L = 10^6 cm^3 = 264.2 US gal = 35.31 ft^3 = 33,814 fl oz
1 bbl = 42 US gal = 5.615 ft^3
1 bbl = 9,702.03 in.^{3}
7,758 bbls = 1 acre-ft = 43,560 ft^3
1 Tcf = .8 million tons per annum (mtpa) of LNG for 20 years

Density

Water = 62.4 lbm/ft^3 = 1000 kg/m^3 = 1 g/cm^3 = 8.33 lbm/US gal
°API, 60°F = (141.5/SG, 60°F) – 131.5
SG, 60°F = 141.5/(°API, 60°F + 131.5)
Low-carbon steel = 490 lbm/ft^3

Velocity

1 knot = 1 nautical mile/hr
60 mph = 88 ft/sec

Temperature

°F = 1.8 (°C) +32
°R = °F + 459.67 = 1.8 (K)
K = °C + 273.18

Energy

1 J = 1 kg·m^2/s^2 = 10^7 dyne-cm = .73756 ft-lbf
1 Btu = 777.9 ft-lbf = 252 cal = 1055 J
1 hp-hr = 2,545 Btu
1 kW-hr = 3,412 Btu = 1.341 hp-hr

Power

1 hp = 550 ft·lbf/s = 33,000 ft·lbf/min = .74570 kW
1 kW = 1000 Watts = 1000 J/s

Gas Constant

R   = 1.987 Btu/lb-mol °R = 1.987 cal/g-mol °K
      = 0.7302 atm·ft^3/lb-mol °R = 1,545 ft·lbf/lb-mol °R
      = 0.08206 L·atm/g-mol K
      = 82.06 atm·cm^3/mol K
      = 8314 Pa·m^3 mol K ox. J/kg mol K
      = 8.314 kJ/kg mol K
      =10.732 psia·ft^3/lb-mol °R

Viscosity

1 cp   = 0.01 Poise = 0.01 g/cm-s = 0.01 dyne s/cm
       = 0.001 kg/m-s = m·P·s = 0.001 N·s/m^2
       = 2.42 lbm/ft-hr = 0.0752 slug/ft-hr
       = 6.72 × 10^{-4} lbm/ft-s
       = 2.09 × 10^{-5} lbf·s/ft^2

$1 \text{ Pa·s} = 0.0209 \text{ lbf·s/ft}^2 = 0.672 \text{ lbm/ft-s}$

$$\left(\frac{\text{lb} \times \text{s}}{\text{ft}^2} \right) = 4.79 \times 10^4 \text{ cp}$$

Constants

$L = 6.022 \times 10^{23} \text{ mol}^{-1} = $ Avogadro's constant

$k = 1.3806 \times 10^{-23} \text{ J/K} = $ Boltzmann's constant

Molar mass of air $= 28.966 \text{ g/mol}$

Standard pressure $= 14.696 \text{ psia}$

Standard temperature $= 60°\text{F}$

Volume of ideal gas $= 379.3 \text{ scf/lb-mol}$ at standard pressure and standard temperature

References and Additional Reading

Adams, N. J. and Charrier, T. 1985. *Drilling Engineering: A Complete Well Planning Approach*. Tulsa, Oklahoma: PennWell Publishing Company.

Ahmed, T. 2006. *Reservoir Engineering Handbook*, third edition. Burlington, Massachusetts: Gulf Professional Publishing/Elsevier.

Allen, T. O. and Roberts, A. P. 1978. *Production Operations: Well Completions, Workover, and Stimulation*, edition. Tulsa, Oklahoma: Oil and Gas Consultants International.

API RP 11L, Design Calculations for Sucker Rod Pumping Systems (Conventional Units), 1988 edition. 1988. Washington, DC: API.

API SPEC 12J, Specification for Oil and Gas Separators, eighth edition. 2008. Washington, DC: API.

Archie, G. E. 1942. The Electrical Resistivity Log as an Aid in Determining Some Reservoir Characteristics. In *Transactions of the AIME*, Vol. 146, No. 1, 54–62, SPE-942054-G. Richardson, Texas: Society of Petroleum Engineers. https://doi.org/10.2118/942054-G.

Arnold, K. E. and Lake, L. W. ed. 2007. *Petroleum Engineering Handbook, Volume 3: Facilities and Construction Engineering*. Richardson, Texas: Society of Petroleum Engineers.

Arps, J. J. 1953. The Effect of Temperature on the Density and Electrical Resistivity of Sodium Chloride Solutions. *J Pet Technol* **5** (10): 17–20. SPE-953327-G. https://doi.org/10.2118/953327-G.

ASME. 2005. *Drilling Fluids Processing Handbook*. Burlington, Massachusetts: Gulf Professional Publishing/Elsevier. https://doi.org/10.1016/B978-0-7506-7775-2.X5000-8.

ASTM D 341, Standard Test Method for Viscosity-Temperature Charts for Liquid Petroleum Products. 2001. West Conshohocken, Pennsylvania: ASTM International.

Baker, R. C. 2000. *Flow Measurement Handbook:, Industrial Designs, Operating Principles, Performance, and Applications*. Cambridge, UK: Cambridge University Press.

Bassiouni, Z. 1994. *Theory, Measurement and Interpretation of Well Logs*, Vol. 4. Richardson, Texas: Textbook Series, Society of Petroleum Engineers.

Bateman, R. M. 1985. *Open-Hole Log Analysis and Formation Evaluation*. Boston, Massachusetts: International Human Resources Development Corporation (IHRDC).

Bell, W. T., Sukup, R. A., and Tariq, S. M. 1995. *Perforating*, Vol. 16. Richardson, Texas: Henry L Doherty Monograph Series, Society of Petroleum Engineers.

Bourgoyne Jr., A. T., Millheim, K. K., Chenevert, M. E. et al. 1986. *Applied Drilling Engineering*, second edition, Vol. 2. Richardson, Texas: Textbook Series, Society of Petroleum Engineers.

Bourgoyne Jr., A. T., Millheim, K. K., Chenevert, M. E. et al. 1986a. *Applied Drilling Engineering*, Vol. II, Sec. 6.4, 290, Fig. 6.5. Richardson, Texas: Textbook Series, Society of Petroleum Engineers.

Bourgoyne Jr., A. T., Millheim, K. K., Chenevert, M. E. et al. 1986b. *Applied Drilling Engineering*, Vol. II, Sec. 1.4, 8, Fig. 1.16. Richardson, Texas: Textbook Series, Society of Petroleum Engineers.

Bourgoyne Jr., A. T., Millheim, K. K., Chenevert, M. E. et al. 1986c. *Applied Drilling Engineering*, Vol. II, Sec. 1.4, 9, Fig. 1.17. Richardson, Texas: Textbook Series, Society of Petroleum Engineers.

Bourgoyne Jr., A. T., Millheim, K. K., Chenevert, M. E. et al. 1986d. *Applied Drilling Engineering*, Vol. II, Sec. 1.5, 13, Fig. 1.23. Richardson, Texas: Textbook Series, Society of Petroleum Engineers.

Bourgoyne Jr., A. T., Millheim, K. K., Chenevert, M. E. et al. 1986e. *Applied Drilling Engineering*, Vol. II, Sec. 4.11, 153, Fig. 4.34. Richardson, Texas: Textbook Series, Society of Petroleum Engineers.

Bourgoyne Jr., A. T., Millheim, K. K., Chenevert, M. E. et al. 1986f. *Applied Drilling Engineering*, Vol. II, Sec. 1.6, 19, Table 1.5. Richardson, Texas: Textbook Series, Society of Petroleum Engineers.

Bourgoyne Jr., A. T., Millheim, K. K., Chenevert, M. E. et al. 1986g. *Applied Drilling Engineering*, Vol. II, Sec. 1.6, 19. Richardson, Texas: Textbook Series, Society of Petroleum Engineers.

Bourgoyne Jr., A. T., Millheim, K. K., Chenevert, M. E. et al. 1986h. *Applied Drilling Engineering*, Vol. II, Sec. 7.1, 302, Fig. 7.1. Richardson, Texas: Textbook Series, Society of Petroleum Engineers.

Bourgoyne Jr., A. T., Millheim, K. K., Chenevert, M. E. et al. 1986i. *Applied Drilling Engineering*, Vol. II, Sec. 5.1, 193, Fig. 5.4A. Richardson, Texas: Textbook Series, Society of Petroleum Engineers.

Bourgoyne Jr., A. T., Millheim, K. K., Chenevert, M. E. et al. 1986j. *Applied Drilling Engineering*, Vol. II, Sec. 2.1, 52, Table 2.7. Richardson, Texas: Textbook Series, Society of Petroleum Engineers.

Bradley, H. B. 1987. Pumping Unit Design Calculations. In *Petroleum Engineering Handbook*, Sec. 10-9, Fig. 10.12. Richardson, Texas: Society of Petroleum Engineers.

Campbell J. M. 1976. *Gas Conditioning and Processing*, fifth edition, Vols. I and II. Norman, Oklahoma: Campbell Petroleum Series, The Campbell Group.

Cholet, H. ed. 2000. *Well Production Practical Handbook*. Paris: IFP Publications, Editions Technip.

Clegg, J. D. and Lake, L. W., ed. 2007. *Petroleum Engineering Handbook, Volume 4: Production Operations Engineering*. Richardson, Texas: Society of Petroleum Engineers.

Crain, E. R. 2018. Special Cases — Coalbed Methane: Sorption Isotherms. In *Crain's Petrophysical Handbook by E. R. (Ross) Crain, PEng*, Chap. 17, https://www.spec2000.net/17-speccbm.htm (accessed 15 February 2018).

Crane Co. 1972. Flow of Fluids Through Valves, Fittings, and Pipe. Technical Paper No. 410, A-24, Engineering Department, Crane Co., Chicago, Illinois.

Dake, L. P. 1978. *Fundamentals of Reservoir Engineering*, No. 8, 151, Fig. 6.4. Amsterdam: Developments in Petroleum Science, Elsevier Science BV.

Drillingformulas.com. 2012. Well Control Equations, http://www.drillingformulas.com/wp-content/uploads/2012/11/Well-Control-Equations-Drillingformulas.pdf (accessed January 2019).

Earlougher, R. A. 1977. *Advances in Well Test Analysis*. Dallas, Texas: AIME.

Economides, M. J. and Nolte, K. G. 2000. *Reservoir Stimulation*, third edition. New York: John Wiley & Sons, Inc.

Ezekwe, N. 2011. *Petroleum Reservoir Engineering Practice*. Boston, Massachusetts: Pearson Education/Prentice Hall.

Fanchi, J. R. and Lake, L. W. ed. 2006. *Petroleum Engineering Handbook, Vol. 1—General Engineering*. Richardson, Texas: Society of Petroleum Engineers.

Fekete Associates Inc. 2014. Material Balance Analysis Theory. http://www.fekete.com/SAN/WebHelp/FeketeHarmony/Harmony_WebHelp/Content/HTML_Files/Reference_Material/Analysis_Method_Theory/Material_Balance_Theory.htm (accessed 04 September 2019).

GPSA. 1974. *Engineering Data Book*, ninth edition, fourth revision. Tulsa, Oklahoma: Gas Processors Suppliers Association.

GPSA. 2004. *GPSA Engineering Data Book*, 12th edition, Sec. 13, Eq. 13-4. Tulsa, Oklahoma: Gas Processors Suppliers Association.

GPSA. 2012. *GPSA Engineering Data Book*, 13th edition, Sec. 3, Fig. 3.1–3.4. Tulsa, Oklahoma: Gas Processors Suppliers Association.

Halliburton Energy Services. 1994. *Log Interpretation Charts*, third edition. Houston, Texas: Halliburton.

Havlena, D. and Odeh, A. S. 1963. The Material Balance as an Equation of a Straight Line. *J Pet Technol* **15** (8): 896–900. SPE-558-PA. https://doi.org/10.2118/559-PA.

Hirschfeldt, M. 2016. API Casing Table Specification (table). OilProduction.net, 18 March 2016, http://oilproduction.net/files/002-apicasing.pdf (accessed 23 January 2019).

Holstein, E. D. and Lake, L. W. ed. 2007. *Petroleum Engineering Handbook, Volume 5: Reservoir Engineering and Petrophysics*. Richardson, Texas: Society of Petroleum Engineers.

Houpeurt, A. 1959. On the Flow of Gases in Porous Media. *Revue de L'Institut Francais du Petrole* XIV (11): 1468–1684.

Hurst, A., Brown, G. C., and Swanson, R. I. 2000. Swanson's 30-40-30 Rule. *AAPG Bull* **84** (12): 1883–1891. https://doi.org/10.1306/8626c70d-173b-11d7-8645000102c1865d.

IADC. 2014. *IADC Drilling Manual*, 12th edition. Houston, Texas: International Association of Drilling Contractors.

Jensen, J. L., Lake, L. W., Corbett, P. W. M. et al. 2000. *Statistics for Petroleum Engineers and Geoscientists*, second edition, No. 2. Amsterdam, Netherlands: Handbook of Petroleum Exploration and Production series, Elsevier B.V.

Krief, M., Garat, J., Stellingwerff, J. et al. 1989. A Petrophysical Interpretation Using the Velocities of P and S Waves (Full-Waveform Sonic). Presented at the 12th International Formation Evaluation Symposium, Paris, France, 24–27 October, Paper HH.

Lake, L. W. ed. 2006. *Petroleum Engineering Handbook, Vol. 2: Drilling Engineering*, Ch. 5, II-223, Fig. 5.2. Richardson, Texas: Textbook Series, Society of Petroleum Engineers.

Lake, L.W. ed. 2007. *Petroleum Engineering Handbook, Volume 7: Indexes and Standards*. Richardson, Texas: Society of Petroleum Engineers.

Lapeyrouse, N. J. 2002. *Calculations for Drilling, Production, and Workover*, second edition. Burlington, Massachusetts: Gulf Professional Publishing/Elsevier.

Lee, J. and Wattenbarger, R. A. 1996. *Gas Reservoir Engineering*. Richardson, Texas: Textbook Series, Society of Petroleum Engineers.

McCain Jr., W. D. 1990a. *The Properties of Petroleum Fluids*, second edition, 330, Fig. 11–13. Tulsa, Oklahoma: PennWell Publishing Company.

McCain Jr., W. D. 1990b. *The Properties of Petroleum Fluids*, second edition, 331, Fig. 11–14. Tulsa, Oklahoma: PennWell Publishing Company.

McCain Jr., W. D. 1990c. *The Properties of Petroleum Fluids*, second edition, 332, Fig. 11–15. Tulsa, Oklahoma: PennWell Publishing Company.

Mitchell, R. F. and Lake, L.W. ed. 2006. *Petroleum Engineering Handbook, Volume 2: Drilling Engineering*. Richardson, Texas: Society of Petroleum Engineers.

Newendorp, P. and Schuyler, J. 2000. *Decision Analysis for Petroleum Exploration*, second edition. Aurora, Colorado: Planning Press.

Penberthy Jr., W. L. and Shaughnessy, C. M. 1992. *Sand Control*. Richardson, Texas: Monograph Series, Society of Petroleum Engineers.

PetroWiki. 2015. Relief valves and relief systems, Fig. 2 (16 July 2015), https://petrowiki.org/Relief_valves_and_relief_systems#cite_note-r2-2 (accessed 11 February 2019).

Rahman, S. S. and Chilingarian, G. V. 1995. *Casing Design, Theory and Practice*. Amsterdam, The Netherlands: Elsevier Science B.V.

Rawlins, E. L. and Schellhardt, M. A. 1935. *Back-Pressure Data on Natural Gas Wells and Their Application to Production Practices*. Baltimore, Michigan: Lord Baltimore Press.

Samuel, R. 2010. *Formulas and Calculations for Drilling Operations*. Salem, Massachusetts: Scrivener Publishing LLC/Wiley & Sons.

Schlumberger. 1969. *Log Interpretation Principles/Applications*. New York: Schlumberger Limited.

Standing, M. B. and Katz, D. L. 1942. Density of Natural Gases. In *Transactions of the AIME*, Vol. 146, No. 1, 144, SPE-942140-G. Richardson, Texas: Society of Petroleum Engineers. https://doi.org/10.2118/942140-G.

Sutton, R. P. 1985. Compressibility Factors for High-Molecular-Weight Reservoir Gases. Paper presented at the SPE Annual Technical Meeting and Exhibition, Las Vegas, Nevada, USA, 22–25 September. SPE-14265-MS. https://doi.org/10.2118/14265-MS.

Thompson, R. S. and Wright, J. D. 1985. *Oil Property Evaluation*, second edition. Golden, Colorado: Thompson-Wright Associates.

Upp, E. L. and LaNasa, P. J. 2002. *Fluid Flow Measurements, A Practical Guide to Accurate Flow Measurement*, second edition. Gulf Professional Publishing/Elsevier.

Warner Jr., H. R. and Lake, L. W. ed. 2007. *Petroleum Engineering Handbook, Volume 6: Emerging and Peripheral Technologies*. Richardson, Texas: Society of Petroleum Engineers.

Willhite, G. P. 1986. *Waterflooding*, Vol. 3, 112, Fig. 4.1. Richardson, Texas: Textbook Series, Society of Petroleum Engineers.

Williams, B. B., Gidley, J. L., and Schechter, R. S. 1979. *Acidizing Fundamentals*, second edition, Monograph Vol. 6. Dallas, Texas: Henry L Doherty Series, SPE of AIME.

Wyllie, M. R. J., Gregory, A. R., and Gardner, L. W. 1956. Elastic Wave Velocities in Heterogeneous and Porous Media. *Geophysics* **21** (1): 41–70. https://doi.org/10.1190/1.1438217.

APPENDIX

Selected pages from the Petroleum Engineering Handbook: *Volume III, p. 346-353.*

TABLE 9.10—BASIC ALLOWABLE STRESS FOR OTHER GRADES OF SEAMLESS PIPE (courtesy of ANSI/ASME)		
Grade	Minimum Temperature	Allowable Stress Minimum Temperature to 100°F
API 5L-A	−20	16,000
API 5LX-42	−20	20,000
API 5LX-46	−20	21,000
API 5LX-52	−20	22,000
ASTM A-106-B	−20	20,000
ASTM A-333-6	−50	20,000
ASTM A-369-FPA	−20	16,000
ASTM A-369-FPB	−20	20,000
ASTM A-524-I	−20	20,000
ASTM A-524-II	−20	18,300

(For additional information, see ANSI *Standard B31.3*, Appendix A.)

• When the variation lasts no more than 10 hours at any one time and not more than 100 hours per year, it is permissible to exceed the pressure rating or the allowable stress for pressure design at the temperature of the increased condition by no more than 33%.

• When the variation lasts no more than 50 hours at any one time and not more than 500 hours per year, it is permissible to exceed the pressure rating or the allowable stress for pressure design at the temperature of the increased condition by not more than 20%.

9.4.4 Wall-Thickness Calculations—Using *B31.4* Code. The ANSI/ASME *Standard B31.4* code is somewhat less stringent than that of *Standard B31.3* because of the lower levels of hazard associated with liquid pipelines. The code for *Standard B31.4* is used often as the standard of design for crude-oil piping systems in facilities, such as pump stations, pigging facilities, measurement and regulation stations, and tank farms. The wall-thickness formula for *Standard B31.4* is stated as

$$ t = \frac{P d_o}{2(F E S_Y)} , \quad \text{..} \quad (9.28) $$

where

t = minimum design wall thickness, in.,
P = internal pressure in pipe, psi,
d_O = OD of pipe, in.,
S_Y = minimum yield stress for pipe, psi (Table 9.11),
F = derating factor, 0.72 for all locations,
and
E = longitudinal weld-joint factor [1.0 seamless, ERW, double submerged arc weld and flash weld; 0.80 electric fusion (arc) weld and electric fusion weld, 0.60 furnace butt weld].

9.4.5 Wall-Thickness Calculations—Using *B31.8* Code. The ANSI/ASME Standard *B31.8* code is less stringent than that of *Standard B31.3,* but more stringent than that of *Standard B13.4.* The *B31.8* code is often used as the standard of design for natural-gas piping systems in facilities, such as compressor stations, gas-treatment facilities, measurement and regulation stations, and tank farms. The *B31.8* wall-thickness formula is stated as

TABLE 9.11—MINIMUM YIELD STRESS AND WELD-JOINT FACTOR (E) FOR COMMON SEAMLESS PIPE MATERIALS (courtesy of ANSI/ASME)

Specification	Grade	Specified Min. Yield Strength, psi (MPa)	Weld Joint Factor E
API 5L	A25	25,000 (172)	1.00
API 5L, ASTM A 53, ASTM A 106	A	30,000 (207)	1.00
API 5L, ASTM A 53, ASTM A 106	B	35,000 (241)	1.00
API 5LU	U80	80,000 (551)	1.00
API 5LU	U100	100,000 (689)	1.00
API 5L	X42	42,000 (289)	1.00
API 5L	X46	46,000 (317)	1.00
API 5L	X52	52,000 (358)	1.00
API 5L	X56	56,000 (386)	1.00
API 5L	X60	60,000 (413)	1.00
API 5L	X65	65,000 (448)	1.00
API 5L	X70	70,000 (482)	1.00
ASTM A 106	C	40,000 (278)	1.00
ASTM A 524	I	35,000 (241)	1.00
ASTM A 524	H	30,000 (207)	1.00
Furnace Butt Welded, Continuous Welded			
ASTM A 53	...	25,000 (172)	0.60
API 5L Classes I and II	A25	25,000 (172)	0.60
Electric Resistance Welded and Electric Flash Welded			
API 5L	A25	25,000 (172)	1.00
API 5L, ASTM A 53, ASTM A 135	A	30,000 (207)	1.00
API 5L, ASTM A 53, ASTM A 135	B	35,000 (241)	1.00
API 5L	X42	42,000 (289)	1.00
API 5L	X46	46,000 (317)	1.00
API 5L	X52	52,000 (358)	1.00
API 5L	X56	56,000 (386)	1.00
API 5L	X60	60,000 (413)	1.00
API 5L	X65	65,000 (448)	1.00
API 5L	X70	70,000 (482)	1.00
API 5LU	U80	80,000 (551)	1.00
API 5LU	U100	100,000 (689)	1.00
Electric Fusion Welded			
ASTM A 134	...	...	0.80
ASTM A 139	A	30,000 (207)	0.80
ASTM A 139	B	35,000 (241)	0.80

$$t = \frac{P d_o}{2 F E T S_Y} \ , \quad \dots\dots\dots\dots\dots\dots\dots\dots\dots\dots\dots\dots \quad (9.29)$$

Specification	Grade	Specified Min. Yield Strength, psi (MPa)	Weld Joint Factor E
TABLE 9.11—MINIMUM YIELD STRESS AND WELD-JOINT FACTOR (E) FOR COMMON SEAMLESS PIPE MATERIALS (continued)			
Electric Fusion Welded (Cont.)			
ASTM A 671	...	Note (1)	1.00 [Notes (2), (3)]
ASTM A 671	...	Note (1)	0.70 [Note (4)]
ASTM A 672	...	Note (1)	1.00 [Notes (2), (3)]
ASTM A 672	...	Note (1)	0.80 [Note (4)]
Submerged Arc Welded			
API 5L	A	30,000 (307)	1.00
API 5L	B	35,000 (241)	1.00
API 5L	X42	42,000 (289)	1.00
API 5L	X46	46,000 (317)	1.00
API 5L	X52	52,000 (358)	1.00
API 5L	X56	56,000 (386)	1.00
API 5L	X60	60,000 (413)	1.00
API 5L	X65	65,000 (448)	1.00
API 5L	X70	70,000 (482)	1.00
API 5LU	U80	80,000 (551)	1.00
API 5LU	U100	100,000 (689)	1.00
ASTM A 381	Y35	35,000 (241)	1.00
ASTM A 381	Y42	42,000 (290)	1.00
ASTM A 381	Y46	46,000 (317)	1.00
ASTM A 381	Y48	48,000 (331)	1.00
ASTM A 381	Y50	50,000 (345)	1.00
ASTM A 381	Y52	52,000 (358)	1.00
ASTM A 381	Y60	60,000 (413)	1.00
ASTM A 381	Y65	65,000 (448)	1.00

NOTES:
(1) See applicable plate specification for yield point and refer to para. 402.3.1 for calculation of S.
(2) Factor applies for Classes 12 22, 32, 42, and 52 only.
(3) Radiography must be performed after heat treatment.
(4) Factor applies for Classes 13, 23, 33, 43, and 53 only.

(For additional information see ANSI/ASME $B31.4$ [7] Table 10)

where
t = minimum design wall thickness, in.,
P = internal pressure in pipe, psi,
d_O = OD of pipe, in.,
S_Y = minimum yield stress for pipe, psi (Table 9.12),
F = design factor (see Table 9.13 and discussion that follows),
E = longitudinal weld-joint factor (Table 9.14),
and
T = temperature derating factor (Table 9.15).

		Type	
Spec. No.	Grade	[Note (1)]	SMYS, psi

TABLE 9.12—SPECIFIED MINIMUM YIELD STRENGTH FOR STEEL PIPE COMMONLY USED IN PIPE SYSTEMS (courtesy of ANSI/ASME)

Spec. No.	Grade	Type [Note (1)]	SMYS, psi
API 5L [Note (2)]	A25	BW, ERW, S	25,000
API 5L [Note (2)]	A	ERW, S, DSA	30,000
API 5L [Note (2)]	B	ERW, S, DSA	35,000
API 5L [Note (2)]	X42	ERW, S, DSA	42,000
API 5L [Note (2)]	X46	ERW, S, DSA	46,000
API 5L [Note (2)]	X52	ERW, S, DSA	52,000
API 5L [Note (2)]	X56	ERW, S, DSA	56,000
API 5L [Note (2)]	X60	ERW, S, DSA	60,000
API 5L [Note (2)]	X65	ERW, S, DSA	65,000
API 5L [Note (2)]	X70	ERW, S, DSA	70,000
API 5L [Note (2)]	X80	ERW, S, DSA	80,000
ASTM A 53	Type F	BW	25,000
ASTM A 53	A	ERW, S	30,000
ASTM A 53	B	ERW, S	35,000
ASTM A 106	A	S	30,000
ASTM A 106	B	S	35,000
ASTM A 106	C	S	40,000
ASTM A 134	...	EFW	[Note (3)]
ASTM A 135	A	ERW	30,000
ASTM A 135	B	ERW	35,000
ASTM A 139	A	EFW	30,000
ASTM A 139	B	EFW	35,000
ASTM A 139	C	EFW	42,000
ASTM A 139	D	EFW	46,000
ASTM A 139	E	EFW	52,000
ASTM A 333	1	S, ERW	30,000
ASTM A 333	3	S, ERW	35,000
ASTM A 333	4	S	35,000
ASTM A 333	6	S, ERW	35,000
ASTM A 333	7	S, ERW	35,000
ASTM A 333	8	S, ERW	75,000
ASTM A 333	9	S, ERW	46,000

The design factor, F, for steel pipe is a construction derating factor dependent upon the location class unit, which is an area that extends 220 yards on either side of the centerline of any continuous 1-mile length of pipeline. Each separate dwelling unit in a multiple-dwelling-unit building is counted as a separate building intended for human occupancy.

To determine the number of buildings intended for human occupancy for an onshore pipeline, lay out a zone ¼-mile wide along the route of the pipeline with the pipeline on the centerline of this zone, and divide the pipeline into random sections 1 mile in length such that

TABLE 9.12—SPECIFIED MINIMUM YIELD STRENGTH FOR STEEL PIPE COMMONLY USED IN PIPE SYSTEMS (continued)

Spec. No.	Grade	Type [Note (1)]	SMYS, psi
ASTM A 381	Class Y-35	DSA	35,000
ASTM A 381	Class Y-42	DSA	42,000
ASTM A 381	Class Y-46	DSA	46,000
ASTM A 381	Class Y-48	DSA	48,000
ASTM A 381	Class Y-50	DSA	50,000
ASTM A 381	Class Y-52	DSA	52,000
ASTM A 381	Class Y-56	DSA	56,000
ASTM A 381	Class Y-60	DSA	60,000
ASTM A 381	Class Y-65	DSA	65,000

GENERAL NOTE:

This table is not complete. For the minimum specified yield strength of other grades and grades in other approved specifications, refer to the particular specification.

NOTES:
(1) Abbreviations: BW–furnace butt-welded; ERW–electric resistance welded; S–Seamless; FW–flash welded; EFW–electric fusion welded; DSA–double submerged-arc welded.
(2) Intermediate grades are available in API 5L.
(3) See applicable plate specification for SMYS.

(For additional information see ANSI/ASME $B31.8$,[9] Appendix D)

the individual lengths will include the maximum number of buildings intended for human occupancy. Count the number of buildings intended for human occupancy within each 1-mile zone. For this purpose, each separate dwelling unit in a multiple-dwelling-unit building is to be counted as a separate building intended for human occupancy.

It is not intended here that a full mile of lower-stress pipeline shall be installed if there are physical barriers or other factors that will limit the further expansion of the more densely populated area to a total distance of less than 1 mile. It is intended, however, that where no such barriers exist, ample allowance shall be made in determining the limits of the lower stress design to provide for probable further development in the area.

When a cluster of buildings intended for human occupancy indicates that a basic mile of pipeline should be identified as a Location Class 2 or Location Class 3, the Location Class 2 or Location Class 3 may be terminated 660 ft from the nearest building in the cluster. For pipelines shorter than 1 mile in length, a location class shall be assigned that is typical of the location class that would be required for 1 mile of pipeline traversing the area.

Location Classes for Design and Construction. *Class 1 Location.* A Class 1 location is any 1-mile section of pipeline that has 10 or fewer buildings intended for human occupancy. This includes areas such as wastelands, deserts, rugged mountains, grazing land, farmland, and sparsely populated areas.

Class 1, Division 1 Location. This is a Class 1 location where the design factor, *F,* of the pipe is greater than 0.72 but equal to or less than 0.80 and which has been hydrostatically tested to 1.25 times the maximum operating pressure. (See Table 9.13 for exceptions to design factor.)

TABLE 9.13—BASIC DESIGN FACTOR (F) FOR STEEL-PIPE CONSTRUCTION OF NATURAL-GAS SERVICE PIPELINES
(courtesy of ANSI/ASME)

| | Location Class | | | | |
| | 1 | | | | |
Facility	Div. 1	Div. 2	2	3	4
Pipelines, mains, and service lines [see para. 840.2(b)]	0.80	0.72	0.60	0.50	0.40
Crossings of roads, railroads without casing:					
(a) Private roads	0.80	0.72	0.60	0.50	0.40
(b) Unimproved public roads	0.60	0.60	0.60	0.50	0.40
(c) Roads, highways, or public streets, with hard surface and railroads	0.60	0.60	0.50	0.50	0.40
Crossings of roads, railroads with casing:					
(a) Private roads	0.80	0.72	0.60	0.50	0.40
(b) Unimproved public roads	0.72	0.72	0.60	0.50	0.40
(c) Roads, highways, or public streets, with hard surface and railroads	0.72	0.72	0.60	0.50	0.40
Parallel encroachment of pipelines and mains on roads and railroads:					
(a) Private roads	0.80	0.72	0.60	0.50	0.40
(b) Unimproved public roads	0.80	0.72	0.60	0.50	0.40
(c) Roads, highways, or public streets, with hard surface and railroads	0.60	0.60	0.60	0.50	0.40
Fabricated assemblies (see para. 841.121)	0.60	0.60	0.60	0.50	0.40
Pipelines on bridges (see para. 841.122)	0.60	0.60	0.60	0.50	0.40
Compressor station piping	0.50	0.50	0.50	0.50	0.40
Near concentration of people in Location Classes 1 and 2 [see para. 840.3(b)]	0.50	0.50	0.50	0.50	0.40

For additional information, see ANSI *Standard B31.8*.

Class 1, Division 2 Location. This is a Class 1 location where the design factor, *F,* of the pipe is equal to or less than 0.72, and which has been tested to 1.1 times the maximum operating pressure.

Class 2 Location. This is any 1-mile section of pipeline that has more than 10 but fewer than 46 buildings intended for human occupancy. This includes fringe areas around cities and towns, industrial areas, and ranch or country estates.

Class 3 Location. This is any 1-mile section of pipeline that has 46 or more buildings intended for human occupancy except when a Class 4 Location prevails. This includes suburban housing developments, shopping centers, residential areas, industrial areas, and other populated areas not meeting Class 4 Location requirements.

Class 4 Location. This is any 1-mile section of pipeline where multistory buildings are prevalent, traffic is heavy or dense, and where there may be numerous other utilities underground. Multistory means four or more floors above ground including the first, or ground, floor. The depth of basements or number of basement floors is immaterial.

In addition to the criteria previously presented, additional consideration must be given to the possible consequences of a failure near a concentration of people, such as that found in a church, school, multiple-dwelling unit, hospital, or recreational area of an organized character

TABLE 9.14—BASIC DESIGN LONGITUDINAL JOINT FACTOR (E) FOR STEEL PIPELINES IN NATURAL-GAS SERVICE
(courtesy of ANSI/ASME)

Spec. No.	Pipe Class	E Factor	Spec. No.	Pipe Class	E Factor
ASTM A 53	Seamless	1.00	ASTM A 671	Electric Fusion Welded	
	Electric Resistance Welded	1.00		Classes 13, 23, 33, 43, 53	0.80
	Furnace Butt Welded—			Classes 12, 22, 32, 42, 52	1.00
	Continuous Weld	0.60			
ASTM A 106	Seamless	1.00	ASTM A 672	Electric Fusion Welded	
ASTM A 134	Electric Fusion Arc Welded	0.80		Classes 13, 23, 33, 43, 53	0.80
ASTM A 135	Electric Resistance Welded	1.00		Classes 12, 22, 32, 42, 52	1.00
ASTM A 139	Electric Fusion Welded	0.80			
ASTM A 211	Spiral Welded Steel Pipe	0.80	API 5L	Seamless	1.00
ASTM A 333	Seamless	1.00		Electric Resistance Welded	1.00
	Electric Resistance Welded	1.00		Electric Flash Welded	1.00
ASTM A 381	Double Submerged-Arc-Welded	1.00		Submerged Arc Welded	1.00
				Furnace Butt Welded	0.60

GENERAL NOTE:

Definitions for the various classes of welded pipe are given in para. 804.243.
For additional information, see ANSI *Standard B31.8*.

TABLE 9.15—BASIC DESIGN TEMPERATURE DERATING FACTORS (T) FOR STEEL PIPELINE IN NATURAL-GAS SERVICE (courtesy of ANSI/ASME)

−20 to 250°F	$T = 1.000$
300°F	$T = 0.967$
350°F	$T = 0.933$
400°F	$T = 0.900$
450°F	$T = 0.867$

For additional information, see ANSI/ASME *Standard B31.8*.[9]

in a Class 1 or 2 location. If the facility is used infrequently, the requirements of the following paragraph need not be applied.

Pipelines near places of public assembly or concentrations of people such as churches, schools, multiple-dwelling-unit buildings, hospitals, or recreational areas of an organized nature in Class 1 and 2 locations shall meet requirements for the Class 3 location.

The concentration of people previously referred to is not intended to include groups fewer than 20 people per instance or location but is intended to cover people in an outside area as well as in a building.

It should be emphasized that location class (1, 2, 3, or 4), as previously described, is the general description of a geographic area having certain characteristics as a basis for prescribing the types of design, construction, and methods of testing to be used in those locations or in areas that are respectively comparable. A numbered location class, such as Location Class 1, refers only to the geography of that location or a similar area and does not necessarily indicate that a design factor of 0.72 will suffice for all construction in that particular location or area (e.g., in Location Class 1, all crossings without casings require a design factor, F, of 0.60).

When classifying locations for the purpose of determining the design factor, F, for the pipeline construction and testing that should be prescribed, due consideration shall be given to the possibility of future development of the area. If at the time of planning a new pipeline this

TABLE 9.16—TYPICAL SURGE FACTORS FOR TWO-PHASE-FLOW PIPELINES (courtesy of API)	
Service	Factor, %
Facility handling primary production from its own platform	20
Facility handling primary production from another platform or remote well in less than 150 ft of water	30
Facility handling primary production from another platform or remote well in greater than 150 ft of water	40
Facility handling gas lifted production from its own platform	40
Facility handling gas lifted production from another platform or remote well	50
For additional information, see API *RP14E*.[10]	

future development appears likely to be sufficient to change the class location, this should be taken into consideration in the design and testing of the proposed pipeline.

9.4.6 Wall-Thickness Calculations—Comparisons. Additional comparison of *Standard B31.3* to both *B31.4* and *B31.8* indicates the following:

• ANSI/ASME *Standard B31.3* is more conservative than either *Standard B31.4* or *B31.8*, especially relative to API 5L, X-grade pipe and electric-resistance-welded (ERW) seam pipe.

• ANSI/ASME *Standard B31.8* does not allow increases for transient conditions.

• The ANSI/ASME *Standard B31.3* specification break occurs at the fence, whereas *B31.8*'s occurs at the "first flange" upstream/downstream of the pipeline.

Using ANSI/ASME *Standard B31.3* criteria for oil- and gas-facility piping will assure a very conservative design. However, the cost associated with the *Standard B31.3* piping design may be substantial compared to the other codes and may not be necessary, especially for on-shore facilities.

9.5 Velocity Considerations

In choosing a line diameter, consideration also has to be given to maximum and minimum velocities. The line should be sized such that the maximum velocity of the fluid does not cause erosion, excess noise, or water hammer. The line should be sized such that the minimum velocity of the fluid prevents surging and keeps the line swept clear of entrained solids and liquids.

API *RP14E*[10] provides typical surge factors that should be considered in designing production piping systems. These are reproduced in Table 9.16.

9.5.1 Liquid-Line Sizing. The liquid velocity can be expressed as

$$V = 0.012\frac{Q_L}{d} , \quad \text{..} (9.30)$$

where

Q_L = fluid-flow rate, B/D

and

d = pipe ID, in.

In piping systems where solids might be present or where water could settle out and create corrosion zones in low spots, a minimum velocity of 3 ft/sec is normally used. A maximum velocity of 15 ft/sec is often used to minimize the possibility of erosion by solids and water hammer caused by quickly closing a valve.

INDEX

Index

Index

www.ingramcontent.com/pod-product-compliance
Lightning Source LLC
Chambersburg PA
CBHW061414210326

41598CB00035B/6214